SPACE 2100

TO MARS AND BEYOND IN THE CENTURY TO COME

+

POPULAR science

SPACE 2100

TO MARS AND BEYOND IN THE CENTURY TO COME

TEXT BY MICHAEL ABRAMS, ERIK BAARD, BRUCE GRIERSON, FENELLA SAUNDERS, WENDY L. SCHULTZ, JEFFREY WINTERS

ORIGINAL ILLUSTRATIONS BY BOB SAULS

ISBN: 1-932273-05-0
Library of Congress Control Number: 2003105223

TIME INC. HOME ENTERTAINMENT
President: Rob Gursha
Vice President, Branded Businesses: David Arfine
Vice President, New Product Development: Richard Fraiman
Executive Director, Marketing Services: Carol Pittard
Director, Retail & Special Sales: Tom Mifsud
Director of Finance: Tricia Griffin
Associate Director: Peter Harper
Prepress Manager: Emily Rabin
Book Production Manager: Jonathan Polsky

Special thanks:
Bozena Bannett, Alex Bliss, Robert Dente, Gina Di Meglio, Anne-Michelle Gallero, Suzanne Janso, Natalie McCrea, Robert Marasco, Mary Jane Rigoroso, Steven Sandonato, Grace Sullivan

POPULAR SCIENCE
Editor in Chief: Scott Mowbray
Deputy Editor: Mark Jannot
Design Director: Dirk Barnett
Science Editor: Dawn Stover
Managing Editor: Jill C. Shomer
Contributing Futurist: Andrew Zolli
Contributing Artist: Bob Sauls

SPACE 2100 was prepared by
Bishop Books, Inc.
611 Broadway
New York, NY 10012

Editorial Director: Morin Bishop
    Project Editor: A. Lee Fjordbotten
    Designer: Barbara Chilenskas
    Assistant Project Editor: Ward Calhoun
    Photography Editor: Alan Gottlieb

We welcome your comments and suggestions about Popular Science Books. Please write to us at:
Popular Science Books
Attention: Book Editors
PO Box 11016
Des Moines, IA 50336-1016

If you would like to order any of our hardcover Collector's Edition books, please call us at 1-800-327-6388 (Monday through Friday, 7:00 a.m.–8:00 p.m., or Saturday, 7:00 a.m.–6:00 p.m., Central Time).

ABOUT THE AUTHORS/ILLUSTRATOR

Michael Abrams is a freelance writer based in New York City who has written for *Discover, Wired,* and *Forbes FYI,* among other publications. He is a master palindromist and co-author of *Dr. Broth and Ollie's Brain-Boggling Search for the Lost Luggage: Across Time and Space in 80 Puzzles.*

Erik Baard is a freelance writer living in New York City who has written for *Popular Science* and the *Village Voice,* among other publications.

Bruce Grierson's science writing appears, with reasonable regularity, in such publications as *The New York Times Magazine* and *Popular Science.* He lives in Vancouver, British Columbia, with his wife, Jennifer Williams.

Bob Sauls is an artist who has created illustrations for virtually every major design project NASA has been associated with in the past 12 years. He has been a contributing artist for *Popular Science* since 1996 and has created some 15 covers for the magazine in addition to numerous illustrations for interior stories.

Fenella Saunders is a science writer/editor at the New York University School of Medicine. Prior to that, she was an associate editor at *Discover.*

Wendy L. Schultz is a futurist engaged in research, writing, teaching, and designing new tools for exploring possible futures. She has facilitated futures thinking on everything from the TransMed pipeline to transhumanists in libraries, everywhere from Pago Pago to Joensuu—but she has not yet been to Mars.

Jeffrey Winters is a writer and editor at *Mechanical Engineering* magazine. His articles about astronomy and physics have appeared in *Discover, Popular Science,* and *The Sciences.*

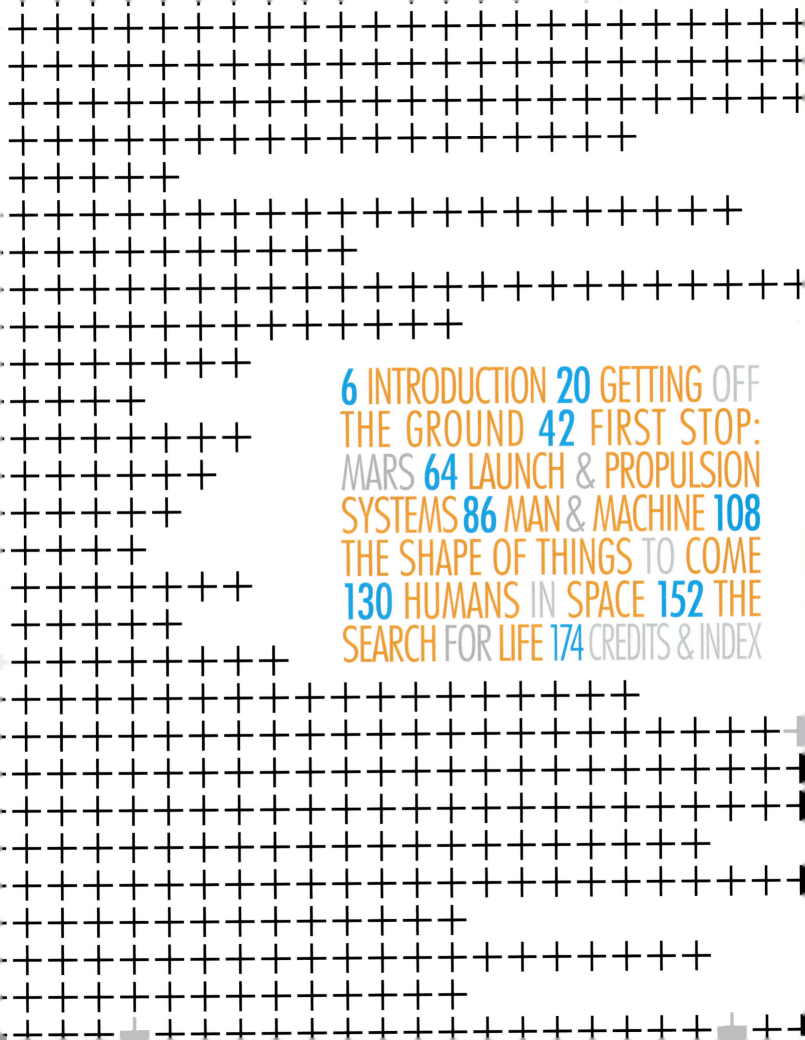

VISION 2100 By Fenella Saunders PREDICTING THE FUTURE IS RISKY BUSINESS. FORTY YEARS AGO, WITH THE COLD WAR SPACE RACE IN FULL SWING, MANY WOULD HAVE PREDICTED THAT BY NOW, WE'D HAVE PERSONAL ROCKET SHIPS AND SPACE COLONIES IN ORBIT, ON THE MOON, AND EVEN MAYBE ON MARS. WELL, THE YEAR 2000 HAS COME AND GONE, AND MANY OF THE MIND-BOGGLING TECHNOLOGIES HAVE FAILED TO MATERIALIZE—NO FLYING CARS, NO SMART PERSONAL ROBOTS, AND NO SPACE

A SLIPPERY SLOPE: The next century may bring another manned moon landing—but this time on Callisto, one of Jupiter's icy moons.

colonies. So a book that aims to predict the future of space travel over the next century might seem to be setting itself up for failure.

But keep this in mind: Even though the present day may not have some of the wildest imaginings that were envisioned half a century ago, we've still made major changes that were not even dreamed of by our parents. The typical "futuristic" computer in a 1960s sci-fi movie is covered with blinking lights and switches and takes up an entire room. There's no sign of a sleek, miniature laptop with crystal-clear graphics. And there was no discussion of other wonders we take as normal today, such as high-resolution digital cameras the size of a deck of cards. Almost all other areas of science and technology have experienced their own revolutions of innovation. We have developed light, strong composite materials, and there's even

ASTRONOMERS ARE OPTIMISTIC THEY'LL FIND AN EARTH-LIKE PLANET IN ANOTHER SOLAR SYSTEM WITHIN A DECADE.

greater promise from entirely new substances such as carbon nanotubes. While we still have a long way to go to solve our power issues, battery technology has advanced significantly. And we are well on our way to controlling our own biology, with a near-complete map of human DNA and rapid developments in genetic engineering. All this has happened in the last 20 or 30 years, and all of these advances could aid in advanced space travel. So maybe it's not so far-fetched that we might actually achieve some of our more bizarre dreams of space exploration in the next 100 years.

In the astrophysics and space technology fields, breakthroughs have also been happening at high speed. Telescopes on Earth have become so good that scientists can use their detailed images to determine the composition of objects millions of miles away. Robotic orbiting spacecraft and landers have provided more in-depth reports on our nearest neighbors. The discovery of planets outside our solar system and anticipated improvements in space imaging technology have made

astronomers optimistic that they'll find an Earth-like planet in another solar system within a decade. This ramping up of new space discoveries has renewed public interest in the chance that extraterrestrial life of some kind might live—or might have lived at some point—somewhere else in the universe. Such public enthusiasm has always provided the momentum needed for giant leaps forward in the space program.

Soon, the national space program won't be the only game in town, either. After all, cars would never have become affordable if only the government had made them. Right now, there are several small aerospace companies hard at work at creating the first private craft that can travel above the Earth's atmosphere and back. In 30 years, it might be commonplace to see throngs of fans gathering around the Los Angeles Spaceport to see if they can catch a glimpse of their favorite Hollywood stars, as they take off and touch down in their own private rocketships. Seeing all this public interest, one of the major airline companies would probably take notice and start their own spacefaring transport, perhaps a Concorde-like craft for elite business travelers that could transport them up into the stratosphere and down to the other side of the globe in under an hour.

Once this sort of transport has proven to be safe, space itself will become the destination, rather than the shortcut. Several companies already have plans on the drawing board for floating hotels in orbit, where everyone could experience the wonders of microgravity. Instead of carting all the pieces up from Earth's surface, space structures like these hotels might be made from molded carbon-based epoxy, injected like toothpaste from a tube into forms in orbit, perhaps over a frame of

THEN AND NOW: British actor Michael Caine (above), as Harry Palmer in the 1967 "Billion Dollar Brain," sits at a typical 1960s sci-fi impression of a supercomputer. A more modern computer display graphs a sequence of the four bases in human DNA (below left)—guanine (G, black line), cytosine (C, blue), thymine (T, red), and adenine (A, green)—and a color display shows the pairing of the bases (below right).

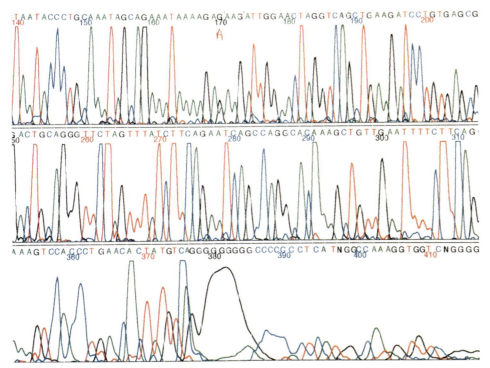

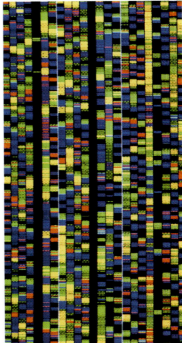

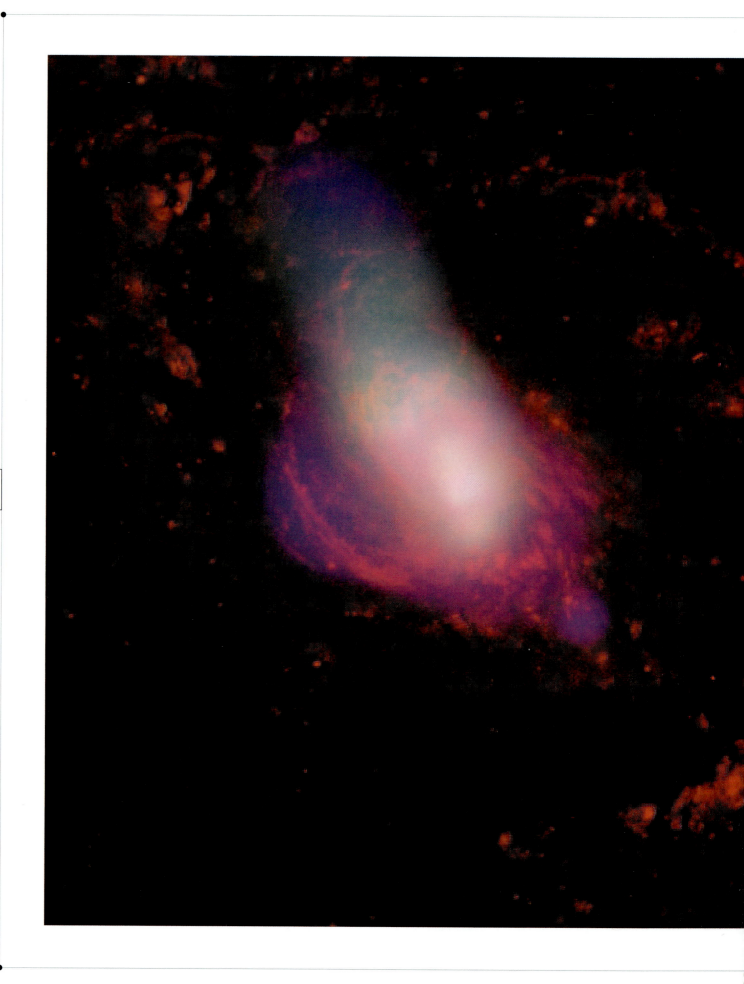

woven carbon nanotubes. With all of the ancillary services that hotels require, a space-based economy would quickly spring up around the orbiting facilities. And with so many people spending so much time in space, by 2050 it's quite likely that babies will not only be conceived in space, but also will be born there. By 2100, as the infrastructure improves, it's possible that people might have spent their entire lives away from our home planet.

For those humans who, around 2050, decide to emigrate from the Earth's surface to a permanent home in space, genetic engineering might make it possible for them to be better equipped for dangers like the extra radiation they'll be constantly exposed to. With the current controversies surrounding genetic engineering, this might seem extreme, but it actually has some historical precedent: Aspiring astronauts in the 1950s and 1960s were willing to put themselves through all manner of extreme physical training on the chance of making it into orbit, in the process modifying their bodies for the rigors of space travel. So there's a good chance that candidates gung-ho to be the first space colonists would not only be willing, but eager, to change their genetic make-up in order to get a chance at the glory of being the first human space resident.

No doubt this will spawn ethical debates on Earth. But in the next few decades, it may become normal to treat diseases with genetic engineering, and in the few decades after that, it could become commonplace to prevent these diseases through DNA manipulation before birth, maybe even before conception. At that point, it would not seem

WHEN THESE HABITATS IN ORBIT ARE ONLY 15 MINUTES FROM EARTH'S SURFACE, WHO WILL HAVE JURISDICTION?

such a stretch to undergo elective genetic treatment in order to increase a person's fitness for extreme environments, such as outer space. We will probably see a paradigm shift in the way people view such genetic medicine by the year 2100, perhaps in no small measure due to our desire to leave the planet's surface.

Another area that will have to undergo public scrutiny is government itself. When all these habitats and businesses in orbit are only 15 minutes away from anywhere on the planet's surface, who will have jurisdiction over them? Who gets to make the laws and collect the taxes? Right now there are treaties in place that make outer space the property of all humankind, but companies won't invest the money to set up hotels and other businesses there unless they can make a profit. By 2030 or so, these issues will have to be ironed out—and no doubt, on issues of economy, governments will be willing to negotiate. Additionally, a modified U.N. or

INTO THE VOID: The Chandra X-Ray Observatory provided this composite x-ray (blue and green) and optical (red) image of the active galaxy NGC 1068, showing gas blowing away in a high-speed wind from a central supermassive black hole. The radiation from such black holes could pose a major hazard for future space travelers.

ONCE WE HAVE A PERMANENT BASE FREE FROM THE PULL OF EARTH'S GRAVITY, TREMENDOUS POSSIBILITIES OPEN UP.

some entirely new international collaborative council will be needed to work out the nitty-gritty details of a new society in space.

Once we have a permanent base free from the pull of our planet's gravity, tremendous possibilities open up for us. Vast arrays of solar cells set up in space, no longer subject to the vagaries of Earth's weather, could have 24-hour exposure to sunlight and produce enormous amounts of power—perhaps even more than needed for orbital structures, allowing the excess to be piped down to Earth's surface as a clean, cheap source of plentiful power. In return, a system of incredibly long, strong tethers, tied to a satellite in orbit, could act as a space elevator and transport goods or pieces of spacecraft cheaply into orbit. Once there, spacecraft could be assembled and powered by something as conceptually simple as a solar sail—an extremely thin sheet of material to catch the solar wind. A fleet of small robotic spaceships could be sent from a space-based station and fly

TOSSING AND TURNING: Adventuresome tourists may soon be able to book very early reservations in a space hotel like this one. With solar collectors on each pod, power would not be a problem, and gravity—to keep one comfortably in bed—would be provided by a gentle, slow rotation of the hotel.

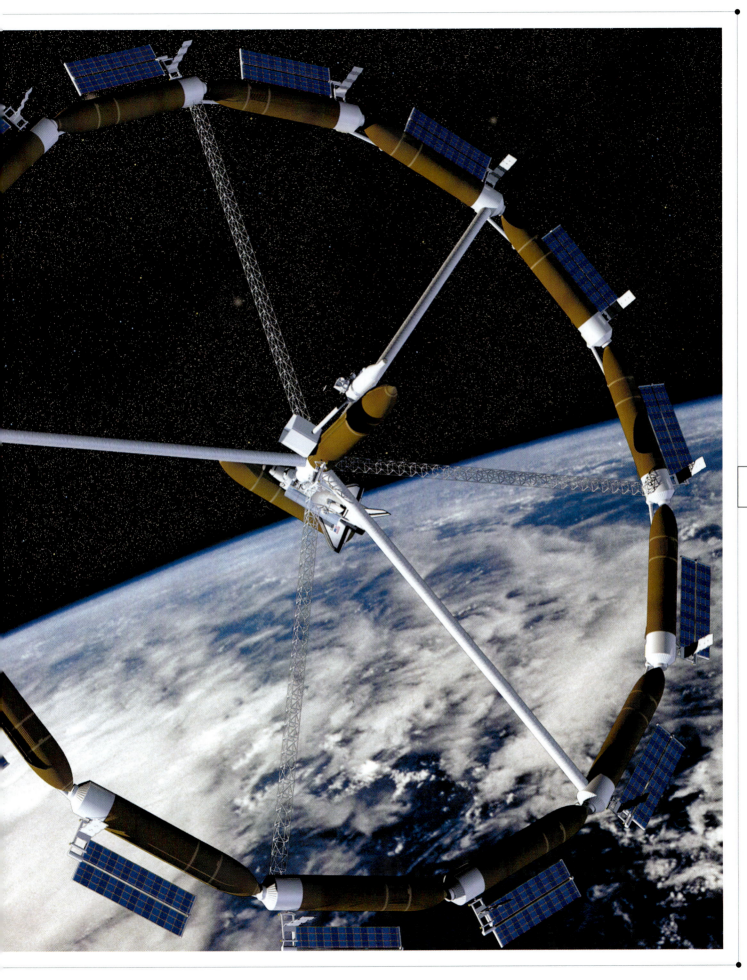

SPACE 2100   INTRODUCTION

via solar sail to survey and mine an asteroid belt, returning the raw materials to space colonists, so they are not as dependent on supplies from Earth. Some NASA designers envision these mining missions happening by 2020.

Learning to use the resources found on-site in space is one of the major requirements for successful colonization of other planets. In fact, before the first human explorers are sent to Mars, NASA's current plans call for a prep team of robotic vehicles to go first and

module could be buried so that it would be shielded from surface radiation. A series of Cryobot robots could be sent to the poles to dig for supplies of liquid water. By the time colonists arrived, they'd be pretty well situated for a long-term stay.

Some NASA documents still call for manned Mars missions by 2020. Soon after that, we might set up long-term research stations on Mars, much like the ones that the National Science Foundation (NSF) has established in Antarctica.

LEARNING TO USE RESOURCES FOUND ON-SITE IN SPACE IS ONE OF THE MAJOR REQUIREMENTS FOR COLONIZATION.

set up everything up for them. A few years ahead of time, a fleet of vehicles would land on the proposed exploration site. Some would immediately start filtering oxygen out of the carbon-dioxide atmosphere and storing it for later human life support, while also taking the filtered carbon from the atmosphere, combining it with hydrogen, and storing the resulting methane as fuel. At first, exploratory teams would use this to fuel their ships for the return flight to Earth, but soon, it might be the main power source for a permanent colony. Other specialized rovers would start preparing the site for colony ships, digging out a pit in which a habitat

Those, in turn, might blossom into cities. But to build the first Levittown in space, we can't rely on habitat modules shipped from Earth. It's very likely that we'd return to the old stand-by, concrete, although in somewhat modified form. Using Lunar regolith, some researchers have shown that the rock can be pulverized to form a pretty good analog of Portland cement. Researchers in Japan have been working on a steam-injection method of curing concrete, using far less water than normal, as H_2O is a rare commodity in space. Spanish researchers say that we should be able to forgo water use altogether in concrete, and use sulfur as a binding agent. To test out these

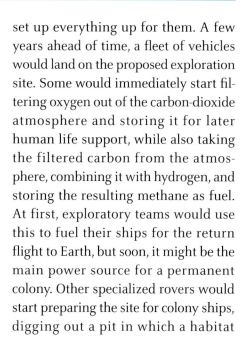

SETTING UP HOUSE:
To set up permanent residence on Mars, we would first establish robotic outposts, which might include a resource utilization system surrounded by solar arrays (above), with rovers assembling the drills for resource recovery (foreground). Looking like children's beach toys playing by themselves, small, lightweight bulldozer robots (right) are intelligent and can work without an operator. As efficient as their bigger brothers, their tiny scoops dig and dump soil into their overhead buckets.

ONCE WE HAVE A FOOTHOLD ON THE RED PLANET, THE REST OF OUR SOLAR SYSTEM MIGHT NOT SEEM SO FAR AWAY.

and other building possibilities, researchers at NASA's Johnson Space Center have developed an ersatz Mars soil made of ground volcanic rock. With all this experimentation under way, Mars and possibly even Moon colonists should arrive with a good repertoire of building methods. There's no reason they couldn't have themselves well established by the turn of the next century.

Once we have a foothold on the Red Planet, the rest of our solar system might not seem quite so far away. Good bets for exploration are Jupiter's icy moons—Europa, Ganymede, and Callisto. These moons show signs of surface ice and possibly subsurface oceans—and liquid water means the possibility of life as we know it. These moons are far colder and more inhospitable to human life than even Mars's poisonous, freezing surface, so by 2100, we may not have permanent outposts there, but it's a good bet that humans will have been sent to take a look around. Using robotic explorers to bore through the surface ice sheet, we could

PARK AND RIDE: Once we begin landing humans on Mars, they will need transportation to and from their work exploring the planet while conserving energy, air, and water. Rovers would need to be designed for a variety of functions.

see if there are volcanic vents in the ocean below—much like the ones that feed deep-dwelling life forms in Earth's oceans, without any exposure to sunlight. And if subsurface life is found on any of Jupiter's moons, by 2100 there would be a clamoring from research scientists to set up a permanent base there in order to do in-depth studies.

If our own solar system proves barren of extraterrestrial life, does that mean that we are alone in the universe? We'll soon have far more places to look. Using only Earth-bound and orbiting telescopes such as Hubble, astronomers say they will soon be able to detect plan-

ets in other solar systems that are of a similar size to Earth and show signs of having an atmosphere potentially hospitable to life. By 2050, if we have space stations or outposts on other planets (away from the interference of our atmosphere and intervening matter), the views of the rest of the universe will get much better. From Mars, it would be possible to send out a space telescope past the asteroid belt near Jupiter, which disrupts viewing beyond our solar system. By the middle of the century, we'll have sent crafts outside of our system as well, and we could map thousands of Earth-like planets for potential visits

FARTHER, DEEPER: When astronauts land on one of Jupiter's moons, they might send a cryobot to penetrate its crust of ice. Once the cryobot reaches the subsurface water, a hydrobot (opposite) would be released to maneuver through the water, analyzing its chemical composition and searching for signs of life. In the century ahead more and more information will come from stronger, better, orbiting telescopes. One of the first of the new breed of telescopes is the Hubble (left), seen here during its second servicing by a Space Shuttle crew in 1997.

from robotic probes. By 2100, we could even be planning to send a human exploration team out of our solar system for the first time, to take a closer look at some of our cousin planets.

The century to come shows all the signs of being a period of major advancement in space travel. All the conditions are right to finally start fulfilling some of the dreams that were promised back in they heyday of the Space Race. By 2100, humans who live in space may have bioengineered cybernetic implants that make their space suits more a part of their bodies than a shell that encases them—they might not even consider

themselves to be exactly the same species as terrestrial humans anymore. The societal changes that accompany such a shift will be dramatic, but it's very likely that we'll take it all in stride—humans are good at adapting that way. It's pretty likely that the next century will reveal other advances so incredible that no one could have even dreamed of them—just as those a century ago would have been hard-pressed to imagine our world today. But even if our dreams are faulty, we must continue to celebrate them, for only a lofty vision of what tomorrow might bring can produce the dazzling future we hope our children will inhabit.

SPACE OUTPOST: A space shuttle docks in this vision of the fully completed International Space Station.

CHAPTER 1

[GETTING OFF THE GROUND]

BY BRUCE GRIERSON

A TRAFFIC ADVISORY PROBABLY SHOULD HAVE BEEN ISSUED FOR LOW-EARTH ORBIT ON APRIL 28, 2001. IT WAS BUMPER-TO-BUMPER UP THERE. A RUSSIAN SOYUZ CAPSULE WITH A GRINNING, CAMERA-TOTING SPACE TOURIST NAMED DENNIS TITO ABOARD APPROACHED THE INTERNATIONAL SPACE STATION (ISS) WITH A VIEW TO DOCKING— BUT IT COULDN'T, BECAUSE THE SPACE SHUTTLE ENDEAVOR WAS ALREADY THERE. BACK ON EARTH, WE MIGHT HAVE BEEN FORGIVEN FOR DUSTING OFF THE OLD PAN AM

SPACE COWBOY: Tito, a California multi-millionaire and former NASA engineer, paid $20 million to visit the space station on April 28, 2001.

First Moon Club card we sent for in 1969 and preparing mentally for the next phase: a little pick-up Quidditch game in the zero-gravity multipurpose room of the Space Hilton, perhaps. After almost a generation of waiting for an encore to the Apollo missions, it looked like space—so tantalizingly close and yet practically so far—was opening grandly before us again.

And then some big things happened in succession, things that snapped our focus back to Earth: a terrorist strike, a hairpin economic down- turn, and a shuttle disaster, all of which left a nation again in mourning and a space station hanging up there half finished. From our portfolios to our kids on the schoolbus, nothing seemed secure.

But crises sometimes force new beginnings. NASA, it's fair to say, is soul-searching to an unprecedented degree. As it tries to define its mission, and to determine to what extent we ourselves will go forth or send robots in our stead, there's no better time to step back and ask the same fundamental questions. What should we be doing in space, and

24

how, and in what order? Who gets to go, and why, and when? With so many terrestrial issues to sort out and sink funds into, should extraterrestrial exploration even continue? The space race was a remnant of the Cold War. We won. We've been to the Moon, left golf balls. Who cares any more?

The answer, as poll after poll indicates, is lots of people. Americans surveyed post-Columbia overwhelmingly wanted a space program to continue. One survey of Americans under 40 revealed that eight of 10 would go up themselves if they could; and fully 10 percent would pay at least a year's salary to do so. Space matters not just because its use is bound up with some of our most pressing terrestrial concerns—from disarmament to environmentalism to campaign finance—but because it is already a part of our daily lives to a degree we sometimes fail to appreciate. We are reminded of that fact when, every once in a while, a pivotal telecommunications satellite goes down, and suddenly cell phones go out and some Internet sites don't come up and you can't get NPR. And you realize just how dependent we are on hunks of metal in geosynchronous orbit that go about their silent business without our ever thinking about them.

Indeed, we are no longer a strictly terrestrial species: That era ended some three years ago, on October 31, 2000, with the beginning of continuous occupation aboard Mir—a milestone relegated in many newspapers to six inches on page A12.

There are valid reasons to spend our money on things other than space. But the best arguments for boldly going . . . well, you know . . . are at least their equal.

EIGHT OUT OF 10 AMERICANS UNDER 40 WOULD GO UP IF THEY COULD; 10 PERCENT WOULD PAY A YEAR'S SALARY TO DO SO.

"There are five reasons for going into space," says Roger Launius, a space historian at the Smithsonian Air and Space Museum in Washington. "We do it for reasons of scientific understanding/advancement, or for national security, or for prestige, or for commercial gain, or for human survival." The last of these is what the German rocket developer Krafft Ehricke identified as the chief point of the "extraterrestrial imperative": to preserve our civilization, to dig out a second beachhead for the species in the event that something happens to this one.

At some point, regardless, we will have to move out of the house, because the Sun, as the British cosmologist Martin Rees puts it, will eventually "lick Earth's face clean." But if that's a couple billion years down the road, do we need to have the car idling in the driveway? Adding up a grim checklist of potential acts of God or man, Rees, a Cassandra with unsettlingly sterling credentials—Astronomer Royal and Fellow of King's College, Cambridge—puts our chances of surviving another

SPACE GOES INTERNATIONAL: Class is in session at the International Space University (top, far left) in Strasbourg, France, where students can get a Master of Space Studies degree; a Chinese taikonaut (far left) prepares for his nation's upcoming first manned foray into space; with the advent of China's space program, including the unmanned Shenzou III (near left), rocketry is returning to its roots.

COFFEE, TEA, GRAVITY?

With the commercialization of space up for grabs, hordes of tourists and would-be entrepreneurs are lining up for a piece of the sky

THERE'S NOTHING LIKE A COLD War adversary to motivate a nation to achieve a grand goal in space. But what happens when international competition dries up? In a post-Apollo, lone superpower world, where does the incentive come from to take space exploration to the next plane?

Never discount the lure of cash. Monetary prizes have always pushed private aviators—from the Wright brothers to Charles Lindbergh to Paul MacCready—to pioneering achievements. The present-day equivalent to Lindbergh's Orteig Prize is the X Prize, a $10-million bounty pulled together by a St. Louis–based foundation to stimulate private ventures in space. The X Prize awaits the first team to build and launch a three-person vehicle

100 km (62.5 miles) into space and back, and repeat the feat two weeks later, thus proving the "reusable" in "reusable launch vehicle."

The X Prize has compelled dozens of companies to develop suborbital

planes, with a view to green pastures just ahead. "There are three potential markets for these vehicles" says Dan DeLong, chief engineer of XCOR Aerospace. "Two of them are credible and one of them is big." The credible: satellite launches from suborbit, and microgravity research of the sort that makes better steel and lenses and drugs. The big one—which many believe holds the key to cheap and reliable access to low-Earth orbit—is space tourism. If the wealthy are willing to go into space for fun, the theory goes, that cash infusion will let the rest of us ultimately get there affordably.

Many believe space tourism will bloom into a multibillion-dollar business. In a recent statement to Congress, Rick Tumlinson of the extraterrestrial-settlement-driven Space Frontier Foundation, called space tourism "the

killer app," "the moneymaker we've all been waiting for." The first wave of adventurers to take a suborbital flight—to experience weightlessness and gawk at the curvature of the Earth on a parabolic roller-coaster ride very similar to that of the Mercury astronauts—will find themselves in rare company. "I'm willing to bet," says video-game designer turned rocketeer John Carmack of Armadillo Aerospace, "that there are at least 500 people willing to pay $100,000 to be among the first." The folks at XCOR whose own planned suborbital, the Xerus, is not an official X Prize candidate) believe at least "three X Prizes' worth of business annually" await the first companies to work out a safe and reliable ride. Indeed, a company called Space Adventures—which brokered the pioneering visits of space tourists Dennis Tito and Mark Shuttleworth to the ISS—is already accepting bookings ("spacecraft for rent, apply now within") aboard suborbital vehicles not yet built. The number of takers will presumably rise as the price comes down. According to a U.S. National Leisure Travel Monitor survey taken in 1997—the first year they included a nonterrestrial question—42 percent of those surveyed said they'd take a trip on a space-cruise vessel, and they'd spend, on average, $10,800 to do it.

Profits from suborbitals will undoubtedly be plowed into the development of orbitals. And once low-Earth orbit (LEO) is humming with traffic, then roadside mechanics, gas-jockeys, and hoteliers will stake their claims. It will be a juncture in space exploration that Werner von Braun seemed to anticipate in 1969. At the press conference following the Apollo 11 moon landing, the great rocketeer leaned in to the microphone and said, "Now we are immortal."

A CROWDED SPACE: More people heading into space will require reusable and hence more economical craft. XCOR's Xerus suborbital space plane (left, above), though not a candidate for the X Prize, is nonetheless one of the most promising projects attempting to meet that goal. Of course, any potential space tourists will have to be prepared for the journey. The group at left, including Hollywood director James Cameron (center), rode the "Vomit Comet," a plane that free-falls to simulate weightlessness.

THERE'S ANOTHER REASON TO BOLDLY GO, OF COURSE. A SIMPLER ONE. WE EXPLORE. IT'S WHAT WE DO.

century at 50-50. "We face a risk no other dominant species has faced: self-extinction by the technologies we have created," says Elon Musk, cofounder of PayPal and president of the private orbital-rocket company SpaceX. "The human race may continue to exist or not. And what does it cost to have an insurance policy? If it's one percent of our annual economy, isn't that money well spent?"

There's another reason to boldly go, of course. A simpler one. We explore. It's what we do. We explore for the same reason a pig wallows in the mud: It's our nature. "This cause of exploration and discovery is not an option we choose," George W. Bush put it, at the memorial service for the Columbia crew, shading forgivably into melodrama: "It is a desire written in the human heart."

And so against this backdrop—now, as luck would have it, on the bicentennial of Lewis and Clark—it's time to look forward again.

We are entering a new era in space in several respects. No longer is it strictly the province of two nations. Students from more than 30 countries now routinely march out of the main

GET OUT OF EARTH'S GRAVITY WELL AND INTO ORBIT, AND YOU'RE HALFWAY TO ANYWHERE IN THE SOLAR SYSTEM.

campus of the International Space University in Strasbourg, France, with a Master of Space Studies degree—thence to find work in space agencies around the world. Within a year, China will have put a taikonaut in space—returning rocketry to its roots. Within a few years, a half-dozen countries will have some sort of independent extraterrestrial presence.

At the same time, a new space race has begun: not between superpowers, but between ambitious entrepreneurs. No longer mere gadflies to the establishment, civilian groups framing an "alternative space agenda" are seizing the day. "Make no mistake," former NASA administrator Dan Goldin told Congress, "it is the private sector that will finally build the machines and provide the access to space to make the dream a reality for all Americans."

The budding small-scale private space industry (as distinct from the big defense contractors) has been likened to the automobile or the aircraft industry circa 1910: We weren't at all sure back then which technology was going to prevail or if it would be cheap enough for the average person to use it. Some-

HARVEST MOON: Mining lunar ice, as in this artist's conception, could yield hydrogen and oxygen to fuel spacecraft or to provide water and air to inhabitants.

30

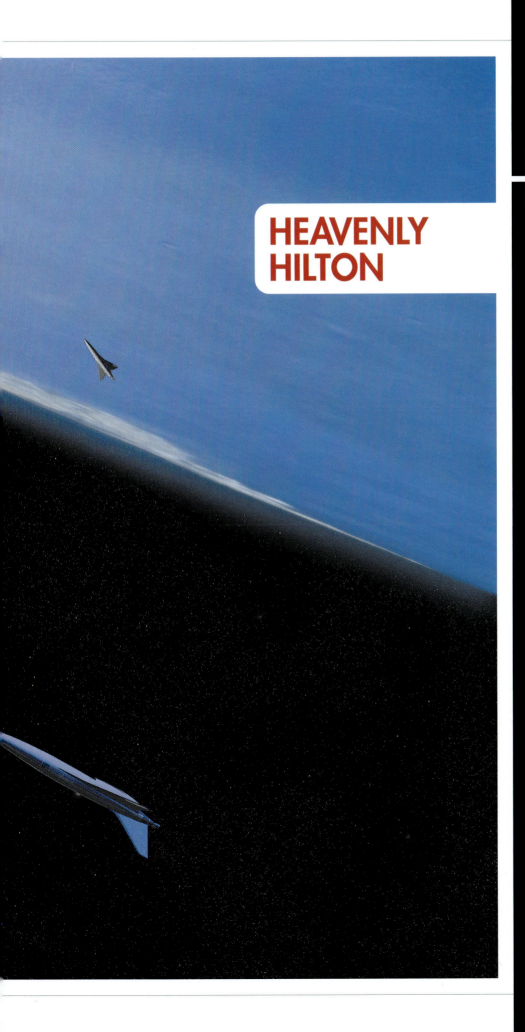

HEAVENLY HILTON

Should space hotels become reality, they will afford space tourists spectacular views of Earth. In order to produce more comfortable and familiar conditions, such exotic establishments will have to spin gently to produce moderate gravity from centrifugal forces. More adventurous tourists may have the opportunity to go to certain areas of the hotel to experience zero gravity.

31

SPACE 2100   GETTING OFF THE GROUND

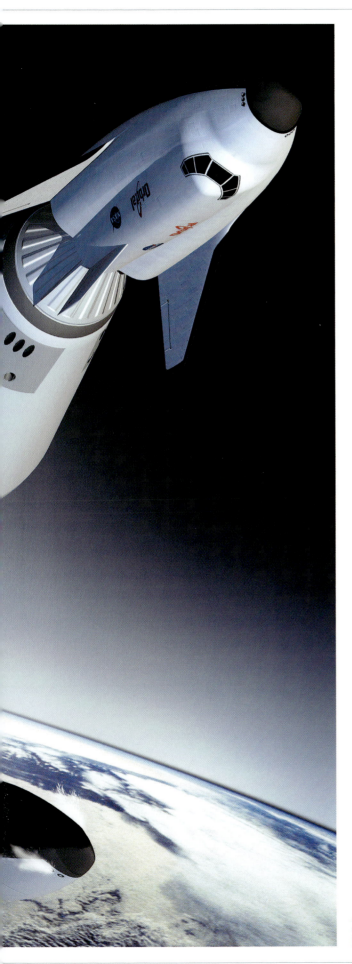

thing big feels imminent. The new astropreneurs are prospectors, often multimillionaire cash-outs from dot-coms or the telecom industry who have gravitated to where the new action is. The rest shave pennies by ordering parts from eBay and pulling scale models behind somebody's souped-up truck in lieu of wind-tunnel testing, all the hungrier for their hardships.

But if real short-run progress in space will indeed come from the private sector, how will that work? The economics of private space development are something of a Catch-22. To attract investment, you have to prove there's a market. To build a market (to create technology cheaply enough), you need investment. No bucks, no Buck Rogers.

Get yourself out of Earth's gravity well and into orbit, said science fiction writer Robert Heinlein, and you're halfway (powerwise) to anywhere in the solar system. As NASA develops its own Orbital Space Plane—a second-generation space shuttle built not for space tourists but to deliver payloads into orbit and ferry astronauts back and forth from the still-under-construction ISS—a speculative landscape of low-Earth orbit (LEO) opens up. Images of space hotels (see page 30) like the one in Stanley Kubrick's *2001: A Space Odyssey* begin to seem feasible. And with market research suggesting that space tourists would be inclined to lay over up there, a couple of orbital hostels are already in various stages of development. Hotel magnate Robert Bigelow has committed $500 million of his own money over 15 years to a floating hotel made of inflatable modules—a project currently under way in the Las

AN EXTRA PUSH: One version of NASA's proposed Orbital Space Plane (OSP) will use an expendable launch vehicle, like Boeing's Delta IV rocket or an Atlas booster, to propel the plane toward its destination.

Vegas desert. Japanese conglomerate Shimizu has long talked up its own orbital hotel and has invested millions in preliminary plans.

Once we're firmly planted in LEO, other things become feasible, such as commercial lunar flybys and Moonshots, the establishment of lunar colonies, and orbital bases for the construction of larger vessels for deep-space exploration. Daunting challenges of physics still thwart the development of alternative energy sources such as lunar-mined helium 3 (the required fusion technology has not been worked out), and solar collector arrays that beam energy to Earth with lasers (the arrays would be of a size approaching cities—an engineering feat and a half). But they dangle as options, a little closer to hand now, depending on how things go back on Earth.

Of course, beyond the technical issues, the wholesale expansion into LEO and beyond is freighted with ethical ones. Can we do this peacefully and cooperatively, as the Outer Space Treaty of 1967 prescribes? Will space remain a kind of common-property resource, or will it get carved up as individual nations and corporations jockey for supremacy—will the paradigm of preservation or development prevail? Will a populace skittish about the dangers of nuclear power ever allow nuclear weapons—land or space-based—to be part of America's ever-evolving strategic defense initiative? If a rocket carrying a nuclear-powered deep-space probe were to crash in its takeoff arc, scattering uranium 238 to the winds, would that be the end of what's still the best hope for planetary travel, nuclear propulsion?

SPACE PIONEERS: Burt Rutan (above) stands in front of SpaceShipOne, the craft he designed to compete for the X Prize. Italian-French mathematician Josef Lagrange discovered five identifiable points in the vicinity of two orbiting masses, such as the Sun and Earth. Points L4 and L5 are located at the apexes of two equilateral triangles that have the two masses at their two vertices. L1 is currently home to the Solar and Heliospheric Observatory Satellite (SOHO), and L2 will soon be home to WMAP. L3, constantly hidden by the sun, is the inspiration for Planet X of science fiction fame.

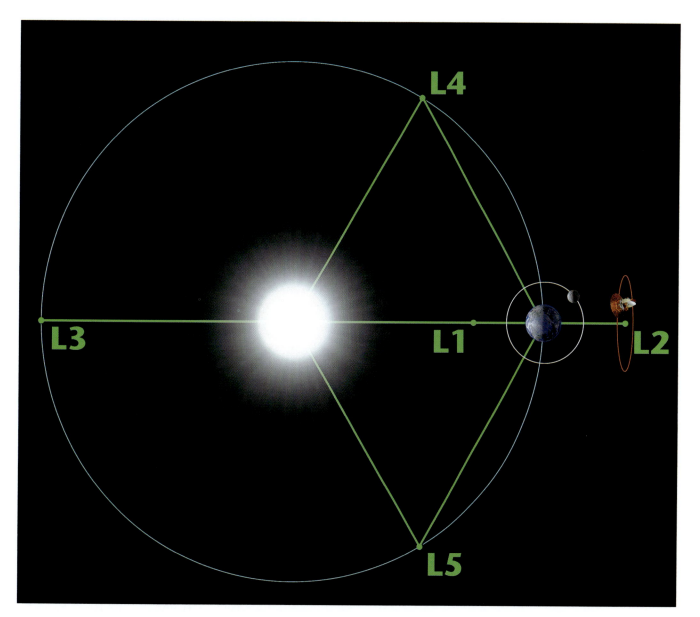

How, then, might the next two decades unfold? Time predictions are in some ways a mug's game, subject to the vagaries of markets, moods, and happenstance. If the Allen Telescope Array in north-central California were to pick up an intelligent signal from deep space, or a space tourist fishing for live extraterrestrial spores from an orbital ketch were to get a strike, well, interest in space would explode—and agendas could turn on a dime. The future of suborbital tourism could depend on a great ad campaign or the enthusiastic return from space of a few high-impact image-makers—the Hollywood producer, the network news anchor, the pro quarterback—who give it the thumbs up. The meme spreads that it's safe up there, and transcendent. Or it doesn't.

But let's assume it does, and the economies of scale that the new market creates solve the problem of cheap access to LEO. What's next? What should the progression of steps, our eventual goals, be? Bases on the orbitally stable Lagrange points L4 and L5? A colony on the Moon, replete with helium 3 mining and light-pollution-free observato-

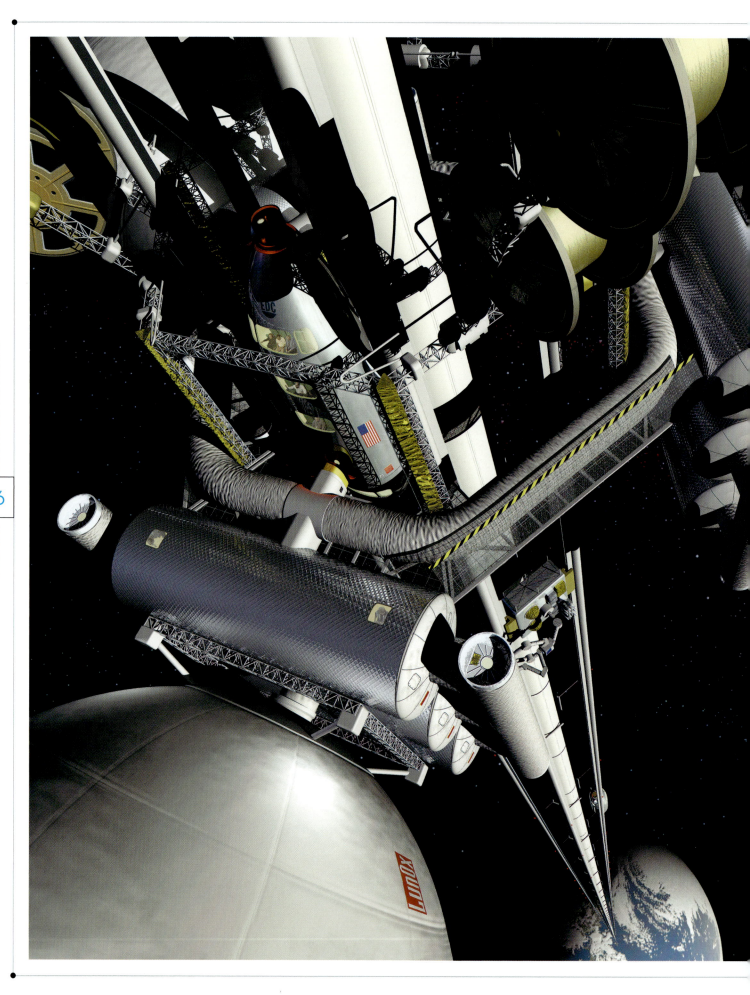

ries and satellite manufacturing plants? A concerted, UN-supported effort to put down roots on Mars? Why or why not? Whatever the answers, the baby steps en route might look something like this:

2004: Recalling the glory days of the '6os, the test pilot of Burt Rutan's small plane called SpaceShipOne, detaches from his ride at 50,000 feet, rockets to an altitude of 62.5 miles over the Mojave Desert, and returns safely to do it again in eight days—thereby capturing the $10-million X Prize. (The money fails to cover the craft's development costs but sparks widespread interest, and heated debate within regulatory agencies. Is it an airplane or a spaceship? The answer has big implications for the ease of similar, future launches.) British and American rovers land on Mars and discover evidence of water. China sends a taikonaut into orbit, becoming the third nation in space.

2005: Space tourists three and four ride a Soyuz rocket to the ISS. Earth's second solar-powered spacecraft is launched from the Barents Sea and, as directed by scientists on Earth, deploys windmill-like blades to tap solar energy and rise into a higher orbit—a harbinger, perhaps, of future deep-space missions.

2006: An American astronaut on a spacewalk cracks a champagne bottle against the newly finished space station. (It will continue to be serviced by the Space Shuttle, NASA announces, until the Orbital Space Plane is finished.) Meanwhile, the first in-space towing company announces operations will soon start up; it will bump wayward satellites into the correct orbit or

THE PRICE OF SUBORBITAL FLIGHTS DROPS WITHIN THE RANGE OF PEOPLE WHO HAVE NEVER READ THE *ROBB REPORT*.

bring them back to the shop for repairs. The first two suborbital space-plane tourist flights leave within days of each other, from Moscow and Mojave. The short parabolic flights cost their passengers—a 16-year-old rock musician from Iceland and a 58-year-old dentist from Tulsa—about $98,000.

2008: Space lotteries, to send civilians of more limited means to the ISS, do roaring business in the U.S. and the U.K.

2009: NASA's SIM (Space Interferometer Mission) probes, launched last year from Cape Canaveral, detect a planet in an Earth-like orbit around a red dwarf star. The planet is dubbed Chomolungma.

2010: NASA's Orbital Space Plane, dubbed Homer and built with the eventual involvement of all three originally competing aerospace contractors, is finished, but without the component of "magnetic launch assistance" many had hoped for. It will be ready for crew rescue in a year, cargo transport in

HOLD THAT ELEVATOR: A space elevator reaching from Earth's surface to geo-stationary Earth orbit (GEO) would provide low-cost transport into space. Immensely strong materials, like carbon nanotubes, are needed to produce the tethers capable of spanning such vast distances.

Rawlings

three. The price of private suborbital flights begins to drop—to around $15,000—which puts them, given the booming economy, within the range of many people who have never read the *Robb Report*.

2011: More people have now been to space than have summitted Mount Everest.

2012: OSP Homer brings its first astronauts to the ISS. Though its original plans called for a launch facility for trips to the Moon and Mars, the space station remains strictly a base for scientific research—albeit one funded now almost exclusively by the U.S. and the E.U. A private company becomes the first entity to successfully launch a small payload into orbit via an electromagnetic railgun track. A Japanese airline reveals that its spaceliner, the Kankoh-Maru, is nearing completion. Meanwhile, laser technology originally developed for the Star Wars program is being repurposed for asteroid defense. A small craft is propelled into suborbit via ice superheated by an array of ground-based lasers.

2013: A private spaceplane skims the lunar surface in a flyby.

2014: China lands a man on the Moon's south pole—rekindling NASA's interests there, if not quite to Cold War levels. Water is found in the subsoil. The Moon is vigorously talked up as a way station, a valuable site for astronomy, construction, helium 3 mining, and payload launches into LEO. China's choice of a landing site—considered the Moon's prime real-estate (the temperature differential between its mountains of eternal light and valleys of eternal shadow is a potential energy source)—causes diplomatic tension. There is confusion over which nation, if any, can lay claim to the Lagrange Points. Private activity is beginning to run ahead of the somewhat murky international laws in this area. Nuclear-propulsion research is stepped up at Sandia and Lawrence Livermore National Labs, amid public debate. Proponents maintain it's still the only feasible technology for long-distance space trips; as for disposal of fissionable material, it may soon be possible to shoot it directly into the Sun. International treaties have thus far proven sufficiently robust to keep nuclear weapons out of space. Talks begin, reasonably seriously, on the feasibility of a small-scale space elevator. These, after chemists at Dupont report a breakthrough in the carbon nanotube technology necessary to produce material strong enough for the Herculean task of ferrying material to orbit.

2015: The two remaining shuttles are retired: There is some nostalgia, though many are glad to see the end of ozone-depleting chlorine injections into the upper atmosphere. Dozens of private companies elbow each other for a spot

PRIVATE ACTIVITY BEGINS TO RUN AHEAD OF THE SOMEWHAT MURKY INTERNATIONAL LAWS IN THIS AREA.

DEFENSE, DEFENSE, DEFENSE: Development of space and all that it promises will necessitate an ongoing defense of our national interests, both in space and at home, with space-based strategies that keep pace with changing technologies. In this artist's conception, an unmanned space maneuver vehicle releases missiles. Given the stakes, it may prove next to impossible to maintain the ideal of a weapon-free space.

A HARSH FRONTIER

Extended stays in space pose challenges that stretch the limits of the human constitution

SPACE IS A HARSH FRONTIER ENVIronment, as the experience of those who have tiptoed into it can attest. About half of shuttle astronauts—a lot not known for delicate constitutions—experience space sickness, and G forces on liftoff are enough to shake your fillings loose. It's a test, but a manageable one.

Beyond short orbital sojourns, how-ever, the stresses and haz-ards of space travel increase dramatically. Over time in microgravity, bones weaken, spines painfully stretch, and cardiovascular systems are taxed. (Experts agree that space hotels, to be habitable by the staff for any length of time, would have to be spun on an axis to generate artificial gravity.) The threat of potentially carcinogenic radiation from cosmic rays and solar flares looms.

And there are psychological factors involved in such ventures as well. The emotional toll of extended confinement, the sense of dis-ease that comes from being even briefly out of sight of Earth, and the strain of personal conflicts ("It could be very difficult to manage a personal relationship that goes sour early in a mission," says Albert Harrison, author of *Spacefaring: The Human Dimension*) are serious issues that have taxed American astronauts and Russian cosmonauts alike.

Deeper space missions pose basic problems of physics as well: how to get ships big enough to carry the required food, fuel, and supplies out of Earth's gravity well? (Constructing them in orbit or on the moon is not without challenge, either.) How to power these missions? How to protect from collisions with space debris? In the case of a Mars mission, what about contamination concerns—from whatever unknown organisms might lurk in the subsoil? Each jump in our progress in the skies—from suborbital to orbital to extraorbital flight—produces exponential increases in complexity.

Optimists will view none of these factors as a showstopper. But the old Boy Scout motto—"Be Prepared"—seems an apt operational dictum.

THE HUMAN TOLL: Monitoring the effects of microgravity—everything from weakened bones to stressed cardiovascular systems—becomes increasingly important as more people venture into space for longer periods of time. Space hotel staff would be especially vulnerable.

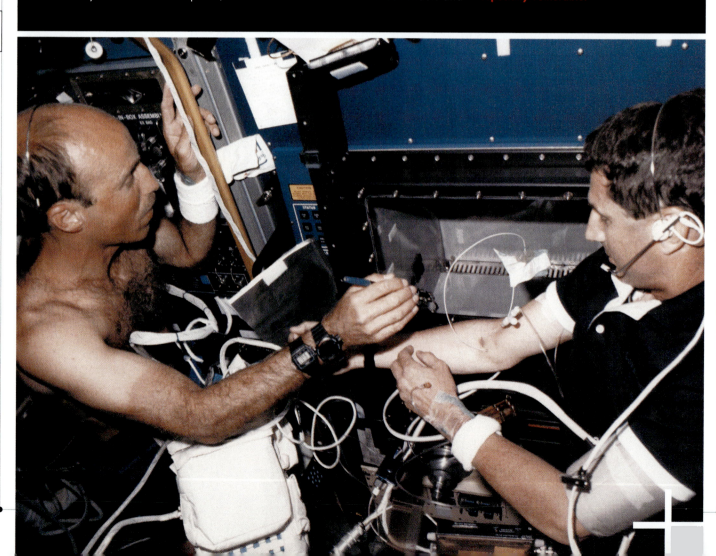

at the pay window of the space-tourism market, which turns over revenues in the tens of billions of dollars. Much of this is plowed back into the development of orbital cruisers.

2017: Chemical analysis of the atmosphere of Earth-like planet Chomolungma, by an interferometric array of telescopes launched the previous year by the European Space Agency, indicates a nitrogen/oxygen/helium mix similar to our own.

2018: The first hypersonic commercial flight leaves New York for Hong Kong, rocketing out of the atmosphere, cresting its parabolic flight over the Midwest and gliding down at Mach 7. Flight time: 55 minutes. Passengers complain that the food is still no great shakes.

2020: When OSP Homer flies to the ISS, it often discharges or picks up a passenger at one of the three private hotel/research stations—one American, one Japanese, and one Russian. These stations are starting to be supplied by thrice-daily care packages launched from Earth by ram cannons. Tertiary orbital business is beginning to take root. An in-orbit satellite assembly outfit has found a niche as telecom executives realize the enormous savings in insurance costs of doing it this way. A New York designer mounts a show of non-tip furniture for use in microgravity.

2023: Space tourists fly a figure eight around the Moon, squeezing off pictures of the Earthrise. There are now space hotels in both elliptical and polar orbits—giving guests the option of different views of the Earth. There are now hundreds of different spaceships tourists can buy tickets on, and one orbital gas station. Breakthroughs

THE FIRST HYPERSONIC FLIGHT LEAVES NEW YORK FOR HONG KONG AT MACH 7. FLIGHT TIME: 55 MINUTES.

in electrothermal propulsion—solar arrays that heat propellants—create fresh impetus to go to Mars *now*. The Mars Direct lobby again makes the case before Congress to go light and go native, manufacturing rocket fuel for the return flight in situ. Spacefarers in Houston and Star City, Russia, begin training for the mission. (Some of the training involves sitting alone in closet-size rooms for days on end.) Congress, hedging the nation's bets, invites bids on orbital construction bays. NASA's Planetary Protection Office (motto: "All the Planets, All the Time"), charged with quarantining and studying extraterrestrial samples brought to Earth, beefs up its staff.

Silver linings? The demise of the Columbia reminded us that the right response in the face of the illimitable cosmos is humility. Spacefaring is still new to us. We are still learning. And so we proceed, boldly but not recklessly. And if we stumble? Then we start again. As Johann Wolfgang von Goethe said, "Life is about failing at greater and greater things."

Throttle up.

DRILLING FOR SOIL: Astronauts take core samples using a portable drilling rig mounted on the back of an unpressurized rover.

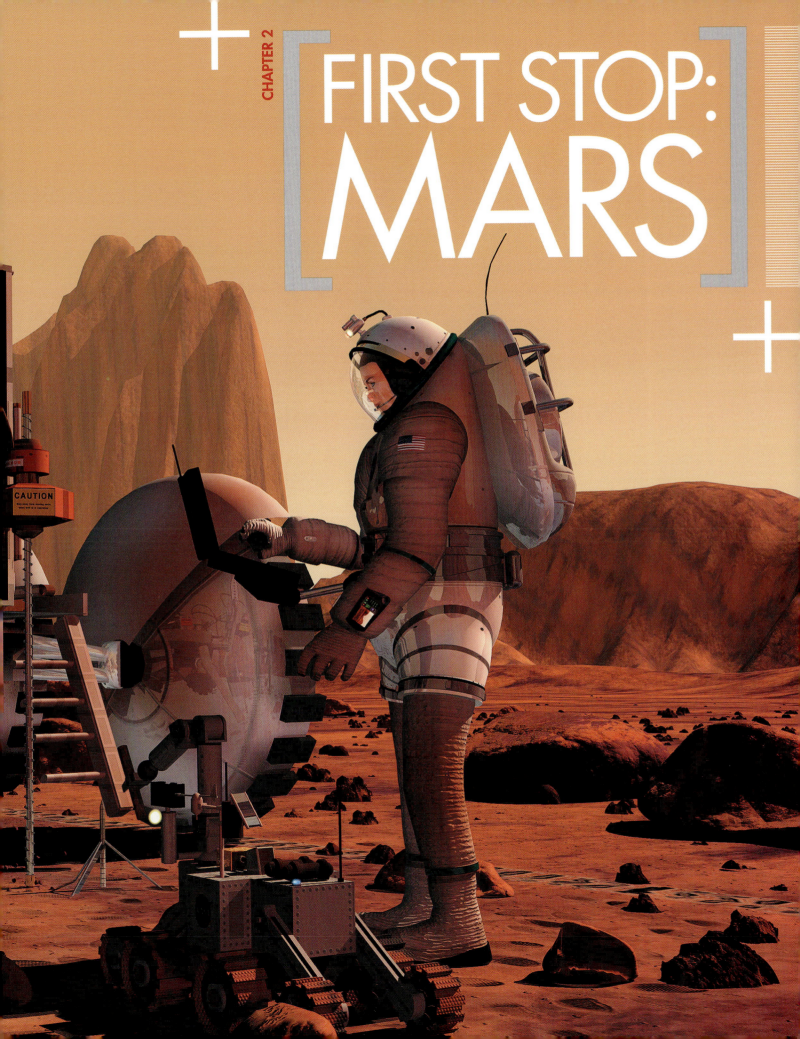

CHAPTER 2

FIRST STOP:
MARS

BY ERIK BAARD AND JEFFREY WINTERS

MARS IS THE FIRST SHORE IN A NEW AGE OF EXPLORATION INSPIRED BY EXTRATERRESTRIAL LIFE, BOTH THE SEARCH FOR IT (DOWN TO MICROBES) AND THE ESTABLISH-MENT OF IT IN THE FORM OF HUMAN COLONIES. MARS COULD ALSO BE THE FIRST PLACE WHERE THOSE TWO DREAMS COLLIDE.

WHY MARS? FAMILIARITY, IN ASTRO-BIOLOGY AT LEAST, BREEDS FASCINA-TION…AND HOPE. NO WORLD HAS INSPIRED MORE OF EACH THAN MARS BECAUSE NO WORLD IN OUR

DUST BOWL: A Mars Global Surveyor photo shows swirling dust storms—caused by temperature contrasts—along the edge of Mars's polar ice cap.

experience looks so similar to Earth, albeit superficially. It's a rocky, or terrestrial, planet formed from the same primordial haze surrounding the Sun as our own world. In terms of size, it falls neatly between Earth and the Moon. And Mars, when compared with Venus, proves the axiom that "less is more." Unlike Venus, which frustrates astronomers with its mirroring clouds, and astrobiologists with the acidic pressure cooker we've discovered beneath its surface below, Mars strikes the casual viewer as offering vistas much like Arizona.

Sure, there could be countless twins for our planet in the cosmos, but the more than 100 worlds we've detected outside of our solar system are all Jupiter-size behemoths. In addition, most are in eccentric inner orbits that would muscle out rocky little planets like ours. It could be decades before new generations of orbiting telescopes find other Earthlike planets with atmospheres bearing the chemical signatures of life.

Planets, of course, aren't the only potential abodes of life. Nearer to home, the moons of

Jupiter tantalize with their shells of ice and hot cores, made volcanic by the stresses of gravitational tides.

But it's on Mars that we long to feel the crunch of soil under our boots as settlers in a rough-hewn territory of steep valleys and rock-strewn flats, and perhaps as prospectors for that most precious of all quarry, life. Much of that ardor can be seen as a cultural legacy, handed down to us for generations spanning 400 years since Galileo Galilei sketched what we'd recognize today as Syrtis Major and, more importantly, used that landmark to clock the rotation of Mars. Our neighboring planet, he correctly deduced, has a north-south axis like Earth's and its days are about the same length as our own. Giovanni Cassini—a competitor whose name is now forever linked with Huygens's in companion probes visiting Saturn and touching down on its moon Titan—confirmed the Dutchman's findings and

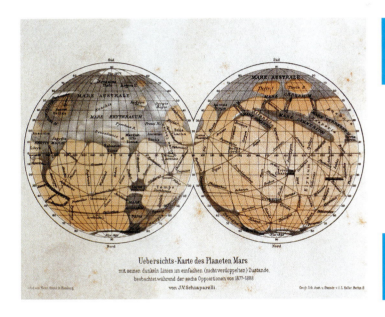

Uebersichts-Karte des Planeten Mars
mit seinen dunkeln Linien im einfachen (nichtverdoppelten) Zustande,
beobachtet während der sechs Oppositionen von 1877–1888
von J.V.Schiaparelli.

BUT IT'S ON MARS THAT WE LONG TO FEEL THE CRUNCH OF SOIL UNDER OUR BOOTS.

MARS MUSINGS: Amateur astronomer Percival Lowell (left) observes Venus during daytime on the 24-inch Clark Telescope in the early 1900s. Lowell was an avid observer of, and theorizer about, Mars. Giovanni Schiaparelli's drawing of Mars and its canals (above), was done between 1877 and 1888 and published in 1898 in Germany. Schiaparelli noticed changing patterns on the Martian surface that we now know are caused by powerful dust storms.

first peered at the heavens through his "optic tube." Telescopic viewings replaced metaphysics in our understanding of Mars, but not always with science. A new mythology arose.

"Love is the child of illusion and the parent of disillusion."
—MIGUEL DE UNAMUNO
(20TH-CENTURY SPANISH PHILOSOPHER)

IT STARTED innocently enough. Two generations after Galileo first saw the disk of Mars in 1609, Italian compatriots, including several Jesuit priests, observed dark patches on its face. Christiaan Huygens, a mathematician from Holland (where the telescope was invented), then added, in 1666, that Mars has white polar caps. His nephew Giacomo Maraldi noticed that those caps are offset from Mars's poles, much like those on Earth, and grow and shrink in what can be regarded as a seasonal rhythm. The renowned English astronomer Sir William Herschel, originally a musician who immigrated from Germany, discovered in the late 18th century that Mars's axis has a tilt of about 25 degrees, just exceeding Earth's 23.5 degrees, further evidence that it shares our planet's experience of seasons. These are very real commonalities, and remarkable science given the technology of the time.

But Mars, as an idea, was already spinning out of control. Herschel argued

that the white polar caps were water ice and snow, and that the dark regions of Mars were oceans, an idea Cassini himself had less forcefully suggested. That notion became the dominant one, with the most notable dissenting view being that the swatches were instead tracts of vegetation. Water "clouds" were thought to be seen, but these were more likely dust. When the possibility of life on Mars was slighted in a popular 17th-century book, Huygens wrote one of his own, *Kosmotheoros*, in which he stated that to suppose other worlds were lifeless would "sink them below the Earth in beauty and dignity; a thing that no reason would permit." He and others supposed that, if anything, the climate of Mars might be milder than that of Earth. That idea was justified by Mars's orientation, but also by how drastically its polar caps diminished.

The true flood of imagination, however, came rushing through canals. Angelo Secchi, director of the observatory at the Collegio Romano and another Italian Jesuit, became fascinated

150 m

SURFACE VIEWS: The 360-degree panoramic shot (above), taken by the Imager for Mars Pathfinder in July 1997, was geometrically improved—the distortion caused by a 2.5-degree tilt of the camera was eliminated, thus removing the curvature in the horizon—and color enhanced. Left: The pattern of branching channels in this February 2003 Mars Global Surveyor Mars Orbiter Camera image suggests sediment and fluid running into a large gulley to the south—at the top of the image—with the slope decreasing from the bottom to the top of the image.

by Syrtis Major during a good viewing period in 1858. He saw the dark blotch—bluish to his eyes—as analogous to a terrestrial ocean, and described it as the Atlantic Channel, or in Italian, *Canale*. The word, now infamous in the annals of astronomy, was officially in play. Others adopted the term, but it was in 1877 that the word was etched into history by Giovanni Schiaparelli, who undertook the first intensive mapping effort of Mars. Rather than broad continents, he saw landmasses sliced apart by *canali*, which can be translated as either "channels" or "canals." It's safe to say he meant the former, because in no way did he ascribe an artificial origin to what he saw. That was left to the most famous, and famously incorrect, Mars observer, American Percival Lowell.

As the 20th century neared, skeptics were grumbling about figments like oceans and canals on Mars. But the wealthy and charismatic American was a compelling writer, and let's face it, he had an appealing story to tell. Even Schiaparelli eventually succumbed to it.

The dark lines he recorded weren't water, Lowell said, but great green oases fed by canals that carried thaw water from the poles of Mars. The speculative tale he wove had at its center a dying world cultivated in such heroic fashion by "intelligent creatures, alike to us in spirit, though not in form."

Lowell's spell faded after his death in 1916, but hopes for life on Mars remained strong. A Soviet scientist proposed in 1959 that one of Mars's two moons, Phobos, is a hollow artificial satellite built by the natives of that planet. We now know that both Phobos and Deimos (Mars's attendants of Fear and Panic in mythology) are captured asteroids that look more like potatoes than Darth Vadar's Death Star. Interestingly, however, Phobos is the closest orbiting moon in the solar system and gets closer to Mars each year. It seems fated to crumble into a ring or crash into the Red Planet in 100 million years or so.

In retrospect, it's easy to mock the life-on-Mars delusion humanity sustained for centuries. After millennia of

it was Herschel, whose optimistic assessment swung open the doors for the canal crowd, who first determined that the atmosphere of Mars was exceedingly thin by the fact that stars passing along the edge of its disk weren't noticeably dimmed.

And perhaps the Mars mania generated by Lowell can be counted among other wrongheaded pursuits that have benefited humanity. Though superstitious, astrology's demand for precise predictions of the heaven's motions bequeathed to us the foundations of astronomy; and the lust for gold through alchemy's experiments gave rise to the science of chemistry. And so the modern obsession with Mars, which may never have existed without Lowell's vision and those of the science fiction writers who followed him, has given us our first real taste of extraterrestrial planetary evolution. Canals aside, Lowell was prescient in his grasp of the impermanence of biospheres. He captured the essence of our understanding—and questions—about Mars today in, of all things, a poem:

. . . the last vestiges of seas
Shall be swallowed up in its cavities,
And Mars like our Moon through space
  shall roll
One waterless waste from pole to pole,
A planet corpse, whence has sped
  the soul.
Already far on with advancing age,
Has it passed its life-bearing age?

IF THE MARS frenzy reached its peak in America, it also collapsed there in 1965. NASA's Mariner 4 flyby probe revealed in 21 close-up photographs a pock-marked and desiccated husk of our preconceived Mars. It was a wholly alien

populating the spiritual universe with gods and angels, the human mind may not have been ready for the challenge posed by an Age of Reason: to stare unblinkingly out into a lonely physical universe. But among those leading science down this long, colorful detour were people who made real contributions to astronomy, and Mars exploration specifically. Schiaparelli brilliantly calculated that the Perseid meteors that shower the Earth in August follow the same orbit around the Sun as the comet Swift-Tuttle, further weaving together our understanding of the solar system into a coherent whole. Herschel discovered Uranus, and

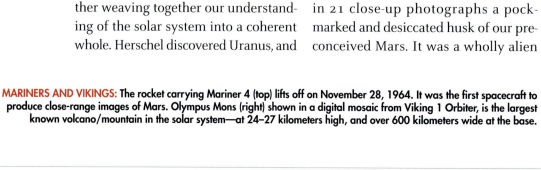

MARINERS AND VIKINGS: The rocket carrying Mariner 4 (top) lifts off on November 28, 1964. It was the first spacecraft to produce close-range images of Mars. Olympus Mons (right) shown in a digital mosaic from Viking 1 Orbiter, is the largest known volcano/mountain in the solar system—at 24–27 kilometers high, and over 600 kilometers wide at the base.

surface, more like the cratered and barren Moon than the living Earth. What spectral data and more conservative telescopic analyses had drummed steadily into academia for years in the first half of the 20th century now arrived in popular culture as a sudden thunderclap: Mars is dead.

Well, at least it was for six years. In 1971, Mariner 9 did for Mars what Lewis and Clark did for North America—opened up vast unknown territories to examination. Most Mars probes from the U.S. and U.S.S.R. had failed miserably, providing only 223 local photographs of Mars since the start of the Space Race. But when Mariner 9 parked itself in orbit, it snapped 7,329 photographs in a working life of over a year. Scientists and science fiction fans alike drooled over this new gallery of the Red Planet: Olympus Mons, the largest volcano in the Solar System towers 24–27 km (79,200–89,100 ft.) over the surrounding plain. The Valles Marineris canyon system at Mars's equator is long enough to bisect North America and cuts as deeply as seven km. The Tharsis region of volcanoes has swelled to a height of 10 km, and at 4,000 km across, it actually warps the sphere of the planet. The southern impact crater Hellas Planitia is over six km deep and 2,000 km wide. And the hardened spreads of lava on Mars bear a striking resemblance to the basalt of the Columbia Plateau of Washington State. Peppering the Red Planet, younger rift valleys, dramatic ridges, rolling hills, and gravelly plains abound. Mariner 9 enchanted Earthlings with a world that was exotic, but not alien.

And scientists viewing those photos also claimed to see telltale signs of water. Lots of it. Was this a case of scientists succumbing to that same old siren song? Decidedly not. If belief in

IT WAS A WHOLLY ALIEN SURFACE, MORE LIKE THE CRATERED, BARREN MOON THAN THE LIVING EARTH.

Lowell's canals stoked the imaginations of H. G. Wells, Arthur C. Clarke, and Ray Bradbury, evidence from the Mariner 9, the Viking landers, Mars Global Surveyor, Mars Pathfinder (and its Sojourner rover), and the recent Mars Odyssey orbiter have served as the basis for a whole generation of doctoral dissertations. There are signs of water in nearly any region you look on Mars. Gullies wrinkle the steep slopes of ridges, and layers of sediment hint at long periods of standing water. Channels—yes, the word is back—that resemble dried riverbeds and paths scoured by flash floods wind across the landscape and often nestle eroded teardrop shaped islands. Even in some flatlands the ground looks to be swollen and cracked by buried ice or collapsed as if water had been emptied from below. And craters often appear to be eroded or ringed by mud piles, perhaps the slurry released from permafrost by the sudden heat of a meteorite impact. Most dramatically, many agree that there was once a great northern ocean into which these rivers

FACTFINDING: The Beagle 2 lander, part of the Mars Express Mission launched by the ESA on June 2, 2003, will gather information on geology, atmosphere, surface environment, the history of water, and the potential for life on Mars.

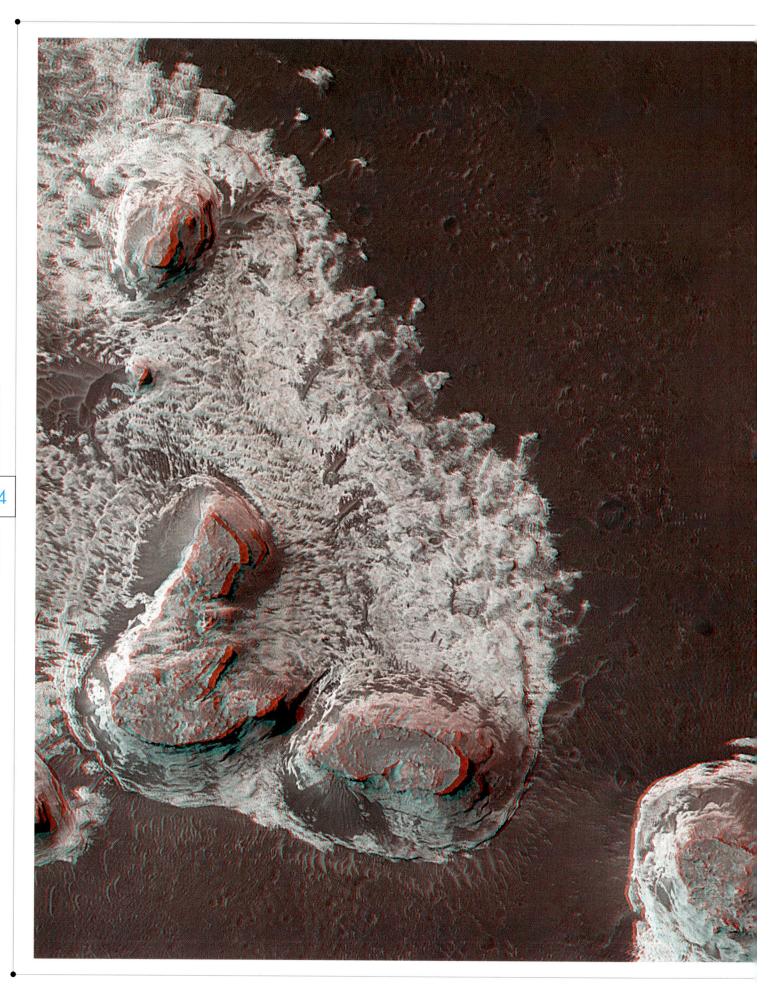

COMING ATTRACTIONS: Meridiani Planum (left), near the Martian equator and prime meridian, is a vast expanse of layered rock. It is the landing site for Opportunity, launched on July 7, 2003, and scheduled to land on January 24, 2004. Aerial Regional-scale Environmental Survey, or ARES (above), is one of four candidates for the Scout Mission, to be launched in 2007. It is designed to be released from a high-altitude balloon, unfold its wings, and fly over Mars, obtaining the first direct measurement of water vapor and chemically active gas concentrations in the near-surface atmosphere of the planet. In a driving test, the Mars Explorer Rover 2 (above right) maneuvers over ramps to test its suspension and range of motion. Spirit and Opportunity, will both carry such a rover to Mars.

drained, and that its shoreline is still discernable.

That apparent distribution is supported by our understanding of how water interacts with gravity in a planet; it famously wants to seek its own level. If ancient oceans sank into the porous ground of Mars, in time it would have seeped and settled evenly around the planet, freezing into a cryosphere that perhaps caps liquid water or is even run through by hot vents and magma.

The new understanding of Mars's water endowment is driving plans for exploring the planet over the next decade or so. The game is now to "follow the water," because that's the most likely place to find life, if it exists on Mars. The water is no longer on the surface, so engineers and scientists are devising novel ways of digging down to levels where it lurks.

A series of probes, set to land beginning in late 2003, is the first to be shaped by this new philosophy. The European Space Agency's Beagle 2 will drop to the surface, cushioned by giant airbags, at Isidis Planitia, a lowland that might be the bed of an ancient northern ocean. There, it will deploy a rock corer and a burrowing device called the

Mole to search for organic molecules—the kind that make up living creatures. It's a short mission, lasting as briefly as 60 days, but there's a chance it might make that historic discovery.

Following close behind (the window for easy travel to Mars opens every two years, so the probes come in waves) are twin NASA rover missions, called Spirit and Opportunity. Opportunity is programmed to land in Meridiani Planum, a region near the equator that possesses a deposit of hematite, an iron-bearing mineral, on Earth generally formed in pools of hot water. Spirit will explore Gusev Crater, which appears to have been a lake at one time. Each rover will carry various spectrometers and cameras designed to determine the composition of the rocks at the landing sites. The hope is to find water ice—or at least evidence of its effect on the local geology.

Even if this armada fails to find clear-cut evidence of water ice, NASA has plans to get data from wider swaths and deeper layers. For example, researchers are working on developing a glider that would enter the Martian atmosphere, unfold long wings, and fly over the surface, looking for gulleys or dry lake beds that would indicate recent flows of water.

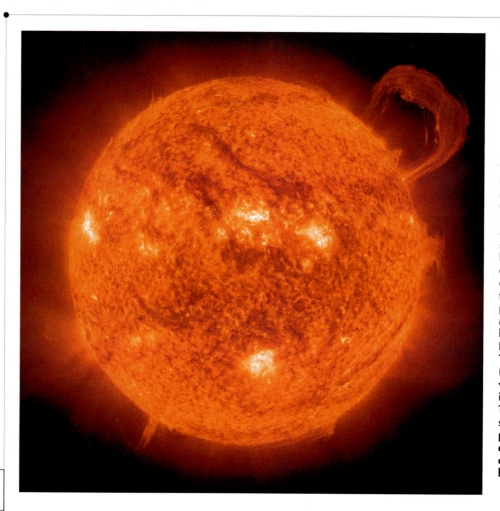

Other missions may drive instrument-laden spears into the ground, blasting several feet below the surface to uncover pristine soil for analysis.

But the mission NASA is aiming for won't launch until 2014 at the earliest. In what would be the most ambitious space mission since the Apollo Moon landings, NASA and its European counterparts would land a several-ton craft on the surface of Mars. Rovers would be dispatched to collect interesting soil samples and deposit them in airtight containers. And then those containers would be launched in a rocket capable of rendezvousing with Earth in 20 months. Such a mission is fraught with complications—and concerns about cross-contamination between the samples and terrestrial life—but those rocks could tell us more after a few weeks in NASA labs than we may learn from three decades of robotic labs in the guts of our Mars rovers.

THE LONG-TERM GOAL is not simply to bring Mars rocks to the scientists, but to send the scientists—and other human explorers—to Mars. Back in the 1960s, this was seen as the next logical step after the Apollo Moon landings. Mars is, after all, the next nearest body that is amenable to a crewed landing. (The surface pressures and temperatures on Venus would quickly destroy any spacecraft.) But budget limitations—and the reality of just what a complex undertaking it would be—forced NASA administrators to shelve those plans.

It's probably a good thing, too. The logistics of a manned mission to Mars are far more complex than those faced by the Apollo program. The Moon is incredibly close by solar system standards—a three-

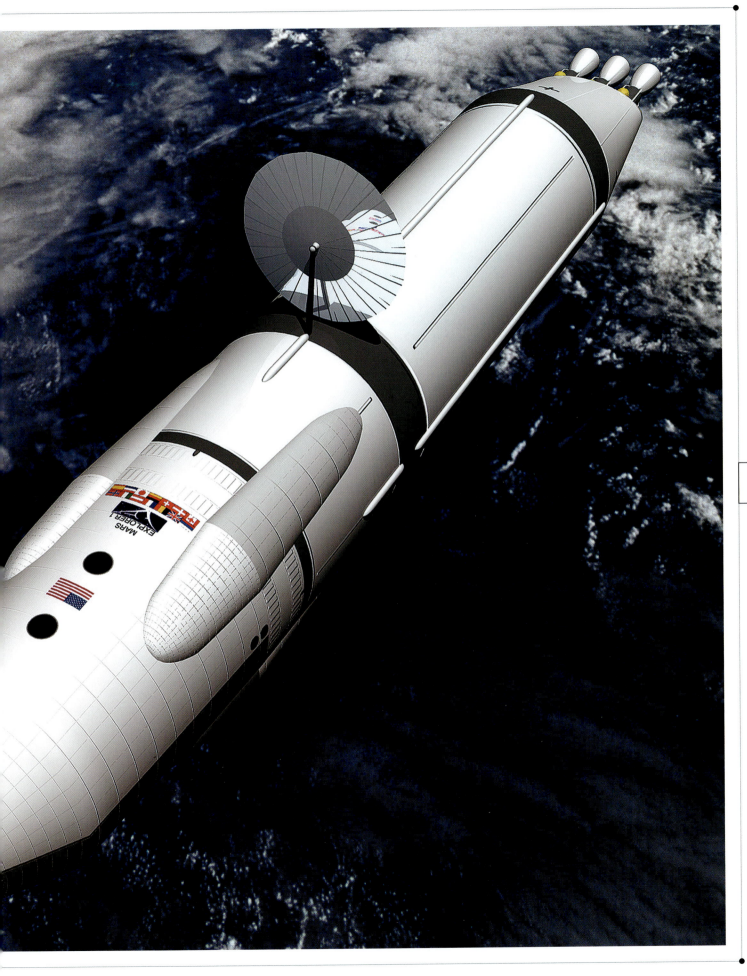

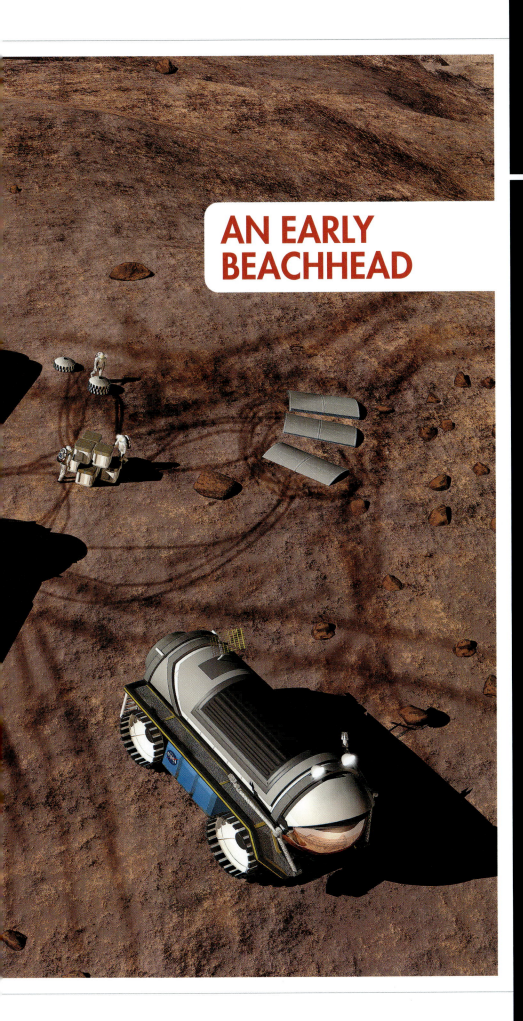

AN EARLY BEACHHEAD

This Mars Base is comprised of four cylindrical modules covered with radiation protection blankets that are filled with Martian soil. The large inflatable structure (upper left) functions as a greenhouse. An inflatable tunnel system with hard connection nodes links the modules. An unpressurized rover for local surface exploration complements a pressurized rover for surface missions of up to two weeks.

day journey. And as the astronauts on Apollo 13 discovered, a spacecraft could make a quick return home if an accident happened en route. That's not a luxury for a Mars mission, where the time in transit is measured in months. Also, no one knows how much physical damage would be done to the astronauts' bodies

documents still refer to a 2020 launch date for a human trip to Mars. Periodically, NASA engineers produce a document outlining the Mars Reference Mission: the step-by-step plan for sending teams to the Red Planet, should the President and Congress give the go-ahead. In past decades, the Reference Mission

SUSSING THE SAMPLES: Astronauts perform on-site analysis of geological material using advanced laser spectrometers. Such on-site analysis will allow astronauts to screen the samples and keep only the most promising ones for further in-depth analysis.

as they lived a year or more with little protection from intense interplanetary radiation. And the psychological stresses of being millions of miles from home, confined with the same small group of people, won't fully be understood until the mission is under way.

Nonetheless, planning for such an adventure still goes on, and many NASA

involved assembling a gigantic spacecraft in low-Earth orbit—similar to the way the International Space Station is being put together. That ship would fly to Mars and remain in orbit, as small, piloted landers would descend to the surface for short excursions. All this made for a cumbersome, $600-billion mission, too expensive for any government to justify.

But in 1990, aerospace engineers Robert Zubrin and David Baker formulated a radically different plan, called Mars Direct. They tossed out the mothership approach and instead proposed that astronauts launched from the Earth could fly to, and land directly on, the surface of Mars. One potential drawback to this approach was how to bring enough fuel to the surface to power the rocket back home. Zubrin and Baker solved this with a brilliant insight: All the makings for rocket fuel are found in Martian surface minerals. Send a pint-size chemical factory ahead of time, they said, and you can make all the fuel you need.

This live-off-the-land approach could be accomplished with off-the-shelf components—no technology need be developed. And this might cost as little as one-twentieth as much as the traditional mission profile. Zubrin and Baker's ideas have had a tremendous influence on the Mars Reference Mission. NASA plans now call for much of the Mars Direct mission profile, from the advance landers to the months-long stay on the surface. But NASA is more technologically ambitious. The Mars Reference Mission still calls for nuclear rockets to propel the piloted ship, and reactors to power the landing site—neither of which has been developed. Even so, it's still a great advance over plans of 15 years ago.

AFTER THEY SAY a little speech and plant a flag, what will astronauts do on Mars? They will probably spend much of their time in search of the same thing their robotic predecessors were looking for: unambiguous signs of life—past or present. With the recent discoveries of abundant water locked up in the Martian crust, life there seems more possible now than it has at any time since the flyby of Mariner 6.

Of course, no one expects to find little green men. At best, life may resemble the "fossils" some scientists believed they found in the Martian meteorite ALH 84001—of primitive single-celled organisms. And even if such organisms still exist, it may be no small feat to discover them. The most likely habitats are deep in the Martian crust, or along volcanic vents that may partially melt the underground ice sheet.

Fossils or other evidence of ancient life may be easier to come by. Although the surface is hostile to organic molecules (the topmost layers of soil are highly toxic), fossils or some kind of

AFTER THEY SAY A LITTLE SPEECH AND PLANT A FLAG, WHAT WILL THE ASTRONAUTS DO ON MARS?

chemical residue of past life forms might be found in the heart of sedimentary rocks formed when the planet had water flowing free across its surface.

By century's end, an explorer might well find another, more advanced form of life on Mars: human colonists. Mars has a surface area equal to the continents of Earth, and scientists and support staff may stay on for long stretches in order to fully explore the place. Long-term outposts, similar to those in Antarctica, will certainly be founded; these might eventually turn into the first extraterrestrial cities.

BY CENTURY'S END, AN EXPLORER MIGHT FIND A MORE ADVANCED FORM OF LIFE ON MARS: HUMAN COLONISTS.

Much of this depends on how hospitable the Martian climate proves to be. The surface is regularly racked by hurricane-force winds, and the thin atmosphere offers little protection from damaging ultraviolet light and other radiation. And of course, what atmosphere there is consists mostly of suffocating carbon dioxide. All this may overwhelm even the most courageous pioneers. But one lesson of human history has been that if people really want to live there, they will find a way.

One way—albeit at the extreme range of possibilities—is a full-scale effort to change the Martian climate— terraforming, it's called. Already, NASA microbiologist Imre Friedmann has identified a microbe, Chroococcidiopsis (see page 160), that might survive in the harsh Martian environment. Over time, as such microbes lived and died, they would change the Martian soil (now mostly dust and rock) and pave the way for more advanced life. It would take centuries to complete, but someone alive today may well live to see the first steps taken toward terraforming Mars.

MOON WITH A VIEW: From one of its moons, Mars is seen during what would be a centuries-long process of converting it to Earthlike climate conditions (terraforming), freeing frozen water and increasing oxygen. A rocket is launching from the moon's surface.

GOING NUCLEAR: A craft like this one powered by pulsed fission could propel us to Jupiter and beyond.

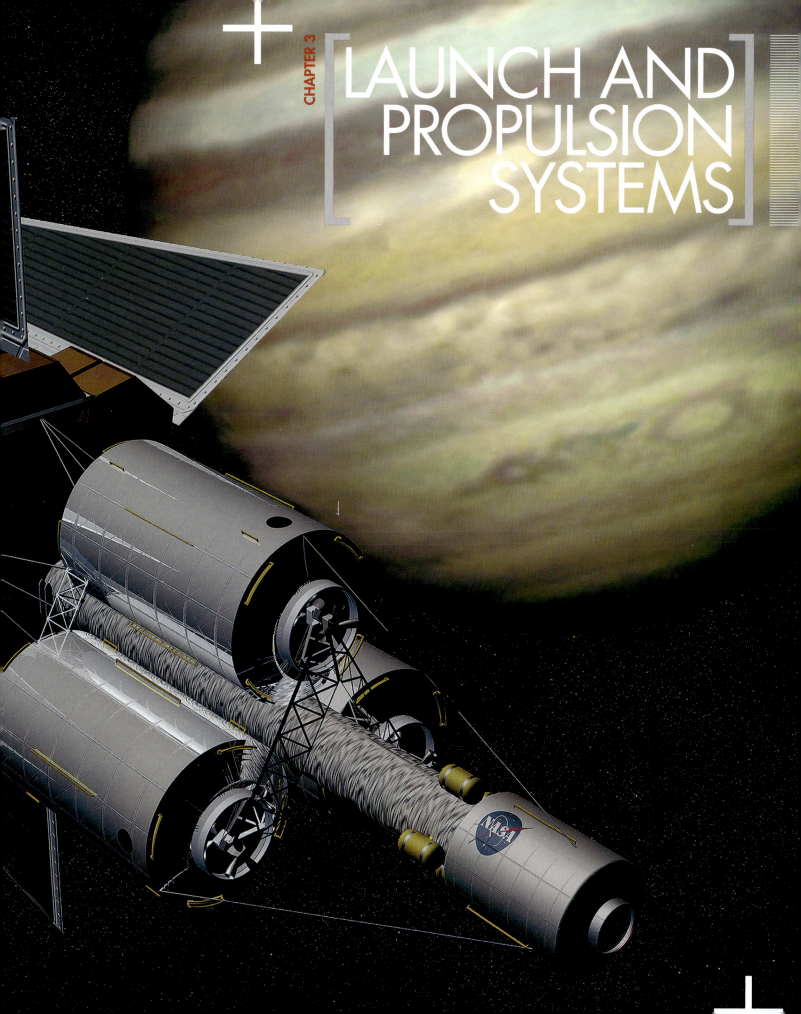

SPACE FLIGHT HAS ALWAYS BEEN DRIVEN BY THE GRANDEST OF ASPIRATIONS: TO WANDER THE STARS. JUST CONSIDER THE VOYAGER SPACECRAFTS, NOW ZOOMING OUT OF OUR SOLAR SYSTEM AT SOME 36,000 MILES PER HOUR. THE SCIENTISTS AND ENGINEERS WHO DESIGNED THE PROBES HELD OUT HOPE THAT THE VOYAGERS MIGHT SOMEDAY ENCOUNTER SOME CREATURE LIVING IN ANOTHER PART OF THE GALAXY, AND SO THEY PACKED ALBUMS OF PHOTOGRAPHS AND

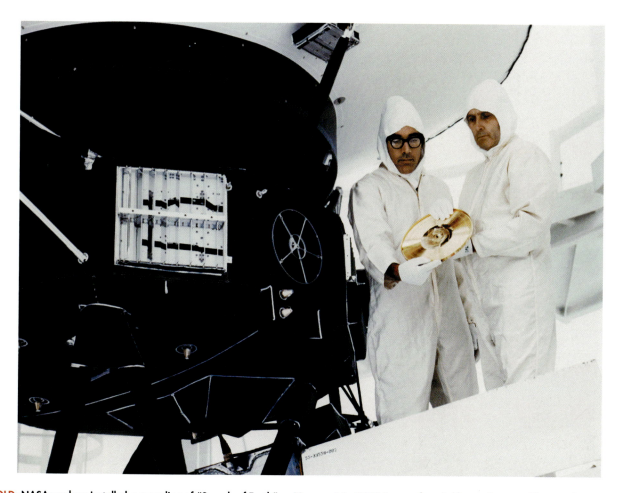

SOLID GOLD: NASA workers installed a recording of "Sounds of Earth" on Voyager 1 in 1977, in case there is life—and a turntable—out there.

attached recordings of songs and greetings from planet Earth.

But for all our dreams, humanity has stayed close to home. The Voyager and Pioneer probes have gone faster and traveled farther than any other manmade objects, but they are still within seven billion miles of Earth. That's a tiny fraction of the distance to the nearest star. At its current rate, it will take Voyager 1 80,000 years to make it as far out as Alpha Centauri, the nearest star system. And it's not headed in that direction.

If that's as fast as we can go, the stars will remain out of reach.

To get a good perspective on the distances involved in interplanetary—and interstellar—travel, you might want to go to Peoria, Illinois. There, the Lakeview Museum has what it calls the world's largest solar system model scattered all over town. The Earth is a four-inch ball inside a gas station; and the Sun is a 36-foot-wide mural on the side of the planetarium a 15-minute walk away. (There isn't a Moon, but if there were, it

would be a ball an inch across about nine feet from Earth.)

On this scale, the orbits of Earth and Mars are only 2,400 feet apart—a six-month trip for an unmanned space probe in the real solar system—and Jupiter's orbit is four miles out. But from there, things begin to expand very quickly. Saturn is some eight miles away (in a Kroger in East Peoria), Pluto is more than 40 miles away, and plaques representing comets and the large icy sub-planets collectively known as Kuiper Belt Objects are scattered across the globe.

What about the nearest stars? On the scale of the Peoria solar system model, Alpha Centauri is 210,000 miles away. That's almost to the Moon. And the center of the Milky Way galaxy would be about where the real Saturn is.

IN THE REAL solar system, traveling by rocket is a lot like getting around the Peoria model by snail. But space travel does offer one small advantage: no friction. So even a small acceleration, applied for a long enough time, could yield a very fast speed. (Whereas a snail is never going to go faster than a few inches an hour.) In theory, if you could accelerate at 1 G (32 feet per second per second) for a year, you could reach nine-tenths the speed of light.

But there's a catch. The more fuel you add to your rocketship, the more slowly it accelerates. Which means you'll have to accelerate longer, which means you'll need more fuel, which means you'll accelerate more slowly still—and on and on. Engineers spell out this paradox in something they call the rocket equation. Simply put, the top speed a rocket can reach (as measured relative to the velocity at which the exhaust

EVEN A SMALL ACCELERATION, APPLIED FOR A LONG ENOUGH TIME, COULD YIELD A VERY FAST SPEED.

shoots out the nozzle) is proportional to the natural logarithm of the percentage of mass left after all the fuel is consumed. Want to go as fast forward as the rocket exhaust exits the rear? You'll need a spacecraft that's 63 percent fuel. Double that speed? Eighty-three percent fuel.

Adding more fuel (or reducing payload) doesn't buy you much more: A rocket that starts off 95 percent fuel can reach only three times exhaust velocity; 99.9 percent will get you to a factor of seven. Even if you could strap the Space Shuttle to a fuel tank the size of the Moon, you couldn't get to Alpha Centauri in less than 6,000 years.

So are we trapped in our solar system? For now, yes. Getting to the far corners of our own system is plagued by the same problems. NASA has proposed a mission to fly by Pluto, but if it ever launches, it will be some eight years in transit. And a combination of the limits of chemical rockets and the quirks of celestial mechanics means that a manned mission to Mars could take 18 to 24 months from lift-off to splash-down.

If we are ever going to follow our

PURPLE HAZE: The clouds of smoke surrounding this launch of the Space Shuttle Columbia in 1996 illustrate the energy needed to escape Earth's gravity well, not to mention the pollution produced in doing so—thus, the search for more efficient and cleaner fuels.

dreams to the stars, we'll need three breakthroughs in launch and propulsion systems. The first two of these—a new way to launch payloads into orbit, and a new type of deep-space propulsion system—are likely to be achieved in the coming decades. But to reach even the nearest stars within a human lifespan will require the discovery of some as-yet-unknown principle of physics.

FROM THE perspective of mission designers, the most expensive 100 miles in the universe are the ones between the ground and low-Earth orbit. Almost every significant use of space is being stymied by the $60,000 a pound it costs the Space Shuttle to lift anything into space. With a price tag that high, the International Space Station is going to be more remote than Antarctica, and space tourism is going to be limited to people with tens of millions of dollars to spare. Longer-term projects—from orbiting power stations to human settlements in space—will remain simply artists' renderings unless the cost to orbit falls dramatically.

There are two main roadblocks to inexpensive access to space: the cost of launch vehicles and the limits of chemical rockets. The favored approach to cutting the cost of the launch vehicle is to reuse all or most of it on multiple flights. But as NASA's experience with the Space Shuttle has shown, repairs and maintenance costs can quickly escalate. And an even more basic problem can be found in the basic rocket equation, which keeps engineers going in circles like a dog chasing its tail. In the end, they wind up with launch vehicles like the Space Shuttle: Fully fueled, it's a 2,000-ton system for delivering 24-ton payloads.

NASA's been trying to find a replace-

THE MOST EXPENSIVE 100 MILES IN THE UNIVERSE ARE THE ONES BETWEEN THE GROUND AND LOW-EARTH ORBIT.

ment for its aging shuttle fleet that could solve at least some of these problems. In the mid-1990s, NASA flight-tested the Delta Clipper: a wingless single-stage vehicle made of lightweight composite material. The Delta Clipper was designed to take-off and land vertically, all without a pilot. Another pilotless design funded by NASA, the Venture Star, looked like an obese space shuttle, but it was intended to deliver payloads to orbit for less than $5,000 a pound and then automatically glide back to a runway.

By 2001, though, funding was dropped for these shuttle replacements. Instead, NASA decided to stick with the decades-old Space Shuttle fleet until at least 2012. In the wake of the Columbia accident in 2003, this now looks to have been a terrible miscalculation.

FRUSTRATED BY the pace of space exploration, entrepreneur Peter Diamandis decided he wanted to spur private development of low-cost, reusable launch vehicles. Taking a page from the early days of aviation, when money prizes were offered for the first flight across the English Channel and the first non-

FAILED PROMISE: The Venture Star, a prototype X-33 Reusable Launch Vehicle, lifts off in this artist's conception, raising hopes for a more reliable and less costly means of putting payload into space. But this option was not to be—the X-33 program was cancelled in 2001.

EYES ON THE PRIZE

With an award of $10 million going to the winner, a plethora of contestants—and approaches—are vying for the X Prize

WANT TO MAKE GETTING INTO orbit really cheap? Find a way to do it without throwing anything away. The Space Shuttle and many of the X Prize entries are reusing their vehicles, and that's a good start. But to turn Earth orbit from an exotic destination to a gateway to the universe, we'll need to get there without throwing away hundreds or thousands of tons of rocket fuel.

NASA researchers currently are looking into propellant-free launch systems, and getting some promising results. In 2000, for instance, the power of an intense laser lifted a five-inch aircraft more than 200 feet into the air. In theory, a much larger laser could lift payloads into low-Earth orbit.

An even more radical idea has been getting some attention recently: Why not build an elevator that *lifts* objects into orbit. No one straps on a rocket pack to get to the top of a skyscraper, after all, so why not take an elevator to the other side of the sky? Some experts believe that this concept may hold the key to making space exploration economically viable.

The space elevator is not a new idea. The Old Testament talks about Jacob's dream of a ladder to heaven, and a 19th-century Russian writer proposed building a tower into space. But the concept didn't attract widespread attention until the 1970s, when Air Force researcher Jerome Pearson published a technical paper showing the feasibility of a space elevator, and Arthur C. Clarke incorporated the idea into his novel *The Fountains of Paradise.*

The key insight behind the space elevator is that the orbital speed of an object circling the Earth depends only

on the location of its center of mass. A satellite in geosynchronous orbit (one that circles the globe in precisely 24 hours, and so remains in a fixed point over the equator) could slowly reel out two lines—one headed down toward the surface, the other away from the Earth—without changing its center of mass or its orbital velocity. If the tethers were long enough, one could dangle from the sky or even be attached to the Earth's surface. Elevator cars could then climb up the tether to low-Earth orbit or geosynchronous orbit, or even slip off the far end of the second tether to be flung into interplanetary space.

A NASA report in 2000 estimated that space elevators could reduce launch costs to less than $10 a pound. This would "open up near-Earth space to miners, explorers, settlers and adventurers," the report concluded. Already, two private companies are laying the groundwork to begin building the first space elevator, though completion is still probably decades in the future, to allow time for development of strong enough materials, like carbon nanotubes.

A more near-term application of the technology is using tethers to transfer satellites from low-Earth orbit to higher ones without fuel. As envisioned by NASA, momentum transfer (or MXER) tethers could swing end-over-end like a giant drum major's baton. As one end swings to its lowest point, it nabs a payload and carries it to the top of the arc, where it is released. In the process, the payload gets a speed boost. A system of these MXER tethers, working in series, could launch a spacecraft toward Mars without burning any fuel in the process.

ECONOMY CLASS: Burt Rutan's SpaceShipOne (above) is paired with White Knight in his effort to win the X Prize and make space travel more efficient and available. Canadian Arrow (left) is a two-stage, three-passenger suborbital rocket also vying for the $10-million award. The low-cost µTORQUE, Microsatellite Tethered Orbit-Raising Qualification Experiment (below), would use a tether from a rocket to toss a microsatellite to the moon.

stop New York-to-Paris flight (the latter claimed by Charles Lindbergh in 1927), Diamandis and members of the St. Louis business community established the X Prize in 1996. The prize—$10 million—can be claimed by the first team to develop and fly a spaceship carrying three passengers to an altitude of 62.5 miles, and then repeat the flight within two weeks.

Maybe it's the sense of competition, maybe it's the cash, but the X Prize has attracted dozens of groups from around the globe. The Aeronautics and Cosmonautics Romanian Association in Ramnicu Valcea, for instance, is building a kerosene-fueled rocket topped by a four-foot-diameter passenger compartment. The Toronto-based da Vinci Project hopes to use a helium balloon to lift its rocket to 80,000 feet before igniting a pair of rockets. And an Argentine group wants to launch an Apollo-type command module it calls the Gaucho.

A competitor that has received a lot of attention is Scaled Composites, run by Burt Rutan, inventor of the Vari-Eze airplane kit and the Voyager aircraft that flew nonstop around the world. His concept involves a Laurel-and-Hardy pair of vehicles, the White Knight and Space-ShipOne. The two craft take off together, with the pudgy SpaceShipOne slung under the White Knight's skeletal, 82-foot wings. After rising to more than 50,000 feet, SpaceShipOne releases and touches off a rocket (burning nitrous oxide and rubber!) that boosts it to 62.5 miles. After three minutes of weightlessness, Space-ShipOne glides back to base.

Other innovative concepts include Kelly Aerospace's winged rocket towed to high altitude by a Boeing 747, and the Ascender, a jet- and rocket-powered space plane being developed by a company in Bristol, England. There seem to be as many plans as there are dreamers,

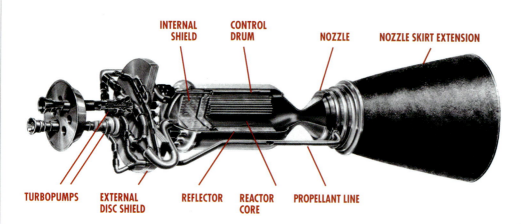

INTERNAL SHIELD CONTROL DRUM NOZZLE NOZZLE SKIRT EXTENSION

TURBOPUMPS EXTERNAL DISC SHIELD REFLECTOR REACTOR CORE PROPELLANT LINE

but no matter who claims the X Prize, this new space race won't really be won until getting to orbit is a normal—and cheap—part of everyday life.

AFTER ALL the hassle of getting into low-Earth orbit, traveling to the rest of the solar system ought to be a piece of cake. In some ways, though, it's just as challenging. The intricacies of celestial mechanics leave no room for error. And the vast distances between planets mean either unleashing enormous amounts of power, or planning for missions that stretch over years—or decades.

Chemical rockets have long been known to have too small a pop to make interplanetary travel anything but a pipedream. Engineers have been tantalized by the promise of harnessing the mightiest of energy sources—nuclear power—to a rocket motor. In the 1950s and 1960s, American scientists were actively developing nuclear propulsion. The Nuclear Engine for Rocket Vehicle Applications (NERVA) project hoped to use an on-board fission reactor to heat fuel that would be shot out the nozzle at a very high velocity. Even further on the edge, Project Orion envisioned riding a wave of minuscule nuclear explosions into orbit and on to Mars. (This really works: A small test rocket that

PULSE POWER: The now-defunct Nuclear Engine for Rocket Vehicle Application (NERVA) thermodynamic engine (above) was intended to produce 75,000 pounds of thrust and potentially reach Mars in as little as 130 days. The multiple mini-explosions proposed by the Orion project seemed even more far-fetched in the 1960s, but with scientists showing renewed interest in nuclear propulsion, the Orion principles could be put into action again to produce a rocket like the one at right.

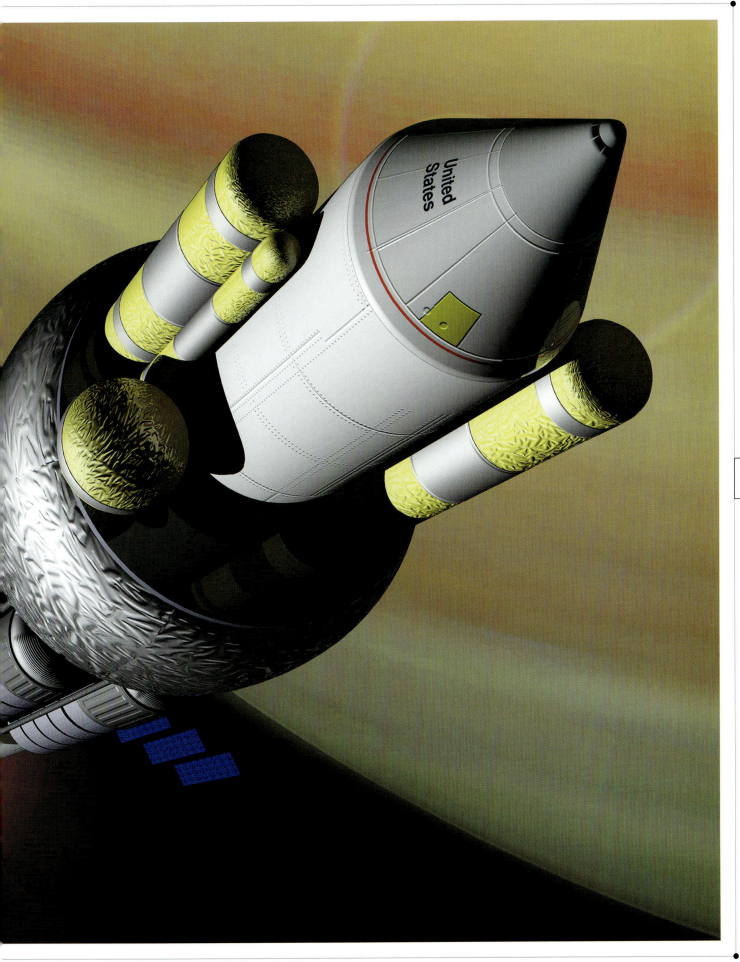

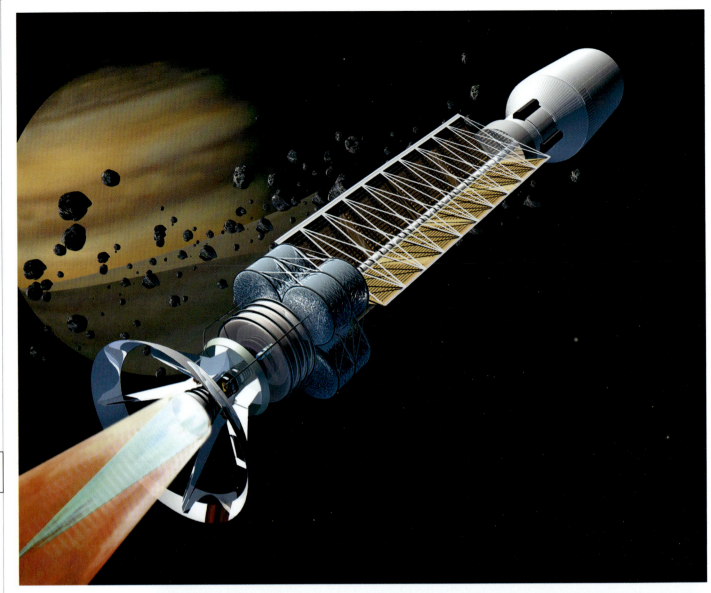

OPPOSITES EXPLODE:
Antimatter is said to be more than 6,000 times as powerful as current nuclear reactors. As a result, it promises the efficiency and power needed to propel craft to Saturn (above) and beyond with no pollution. Once the stuff of comic books, antimatter is real, but incredibly expensive to manufacture and difficult to contain. Theoretically the antimatter would combine with matter in an engine (right) and the resulting annihilation process would produce violent bursts of light and energy.

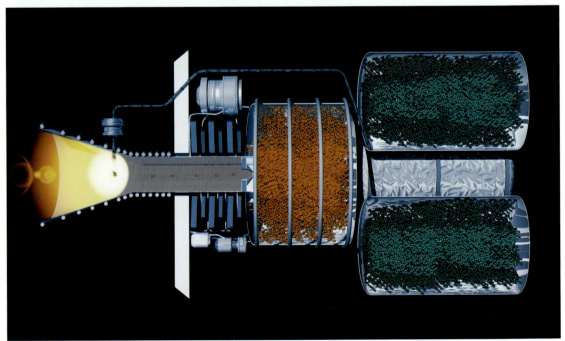

used packets of conventional explosives rather than nukes was launched from the ground to an altitude of 200 feet.)

Concerns over radiation—and the fear of a catastrophe involving a rocket full of nuclear bombs—killed nuclear rockets in the mid-1960s. But a new generation of engineers is reviving interest in nuclear-powered craft, though only for deep-space flight. NASA's Project Prometheus is designing fission reactors for powering electric propulsion systems or even directly heating propellant. And in 1997, when a NASA team drew up a plan to send astronauts to Mars and back, they proposed using technology almost identical to the NERVA. Such a nuclear rocket could reach Mars in as little as 130 days.

Other researchers are looking into powering rockets with an even more energetic reaction—matter-antimatter annihilation. Penn State physicist Gerald Smith has proposed shooting microscopic flecks of antimatter into hydrogen-uranium pellets and using the intense heat from annihilation to touch off a fusion reaction in the rocket nozzle. The system, called ICAN II, would be incredibly powerful—a trip to Mars could be cut to just 45 days. But even the modest amounts of antimatter needed for a one-way mission (less than a millionth of a gram) would be time-consuming to produce and tricky to contain. If the magnet bottle holding the antimatter was turned off for just a fraction of a second, the spacecraft might be destroyed.

IN SPACE, you don't always need a big bang to reach high speeds. Because there's no drag, a slow, steady acceleration can work wonders. That's the concept behind electric drives, which shoot out atoms one or a few at a time at very high speed. No individual atom adds

A NEW GENERATION OF ENGINEERS IS REVIVING INTEREST IN NUCLEAR-POWERED CRAFT FOR DEEP-SPACE FLIGHT.

much acceleration. But electric drives are incredibly fuel efficient and can be left on long enough for those minuscule pushes to add up to quite a wallop.

Because of that trade-off—low thrust for high efficiency—electric drives can't lift a spacecraft into orbit. NASA launched Deep Space 1 with a conventional rocket, then switched on an electric drive that sent the probe to a close observation of Comet Borrelly. The mission was so successful that NASA scientists are considering using electric drives to power future unmanned missions.

Manned missions will need something a little more powerful than the motor on Deep Space 1. Plasma physicist and astronaut Franklin Chang-Diaz is leading a NASA team to develop an electric drive concept called the Variable Specific Impulse Magnetoplasma Rocket (VASIMR). This drive uses a device like a microwave oven to heat a gas to some 10 million degrees, then accelerates it through a magnetic field to speeds as high as 100,000 miles per hour. The amount of plasma shooting out the end can be controlled, to provide either greater acceleration or fuel efficiency.

Such a system could provide both a

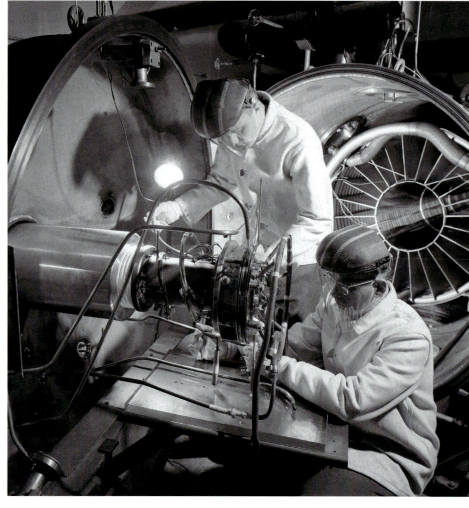

ELECTRIC POWER: The first ion engine (top left, in 1959) was a predecessor to the system used to power Deep Space I in its exploration of the solar system (far right). The solar-powered ion propulsion engine (top right) used on the mission was the first non-chemical system to be used as the primary means of propelling a spacecraft.

powerful surge to get out of Earth orbit and slow-but-steady acceleration across interstellar space. But VASIMR would be a power hog, requiring a 12 megawatt nuclear reactor to keep the plasma flowing. Until humans decide to accept the risk of launching that much nuclear material into space, VASIMR will be limited to small-scale tests.

ROCKETS AND other craft are often whimsically referred to as space "ships." But some engineers want to turn this phrase into a reality by rigging spacecraft to enormous sheets and literally sailing them through the sky. It sounds fanciful, but it's an idea grounded in physics. As sunlight reflects off the surface of a shiny bit of metal, it exchanges the tiniest bit of momentum with the object. Hoist a large enough metal sheet—say

a few miles across and a few millionths of an inch thick—and the momentum can build up over time to interplanetary speeds.

It's an old idea: Reportedly Johannes Kepler wrote about it to Galileo, and Russian Konstantin Tsiolkovsky worked out the physics in the 1920s. But it hasn't been feasible to construct such filmy sails until recently, when superstrong plastics such as Mylar were developed. But small experiments, especially the Russian Znamia, deployed in 1993, have proved the concept. And in theory, a large sail ship could explore the outer reaches of the solar system.

Of course, as you travel farther out into the solar system, the intensity of sunlight diminishes in proportion to the square of the distance. To make sails really effective, you need a tight beam of intense

ELECTRIC DRIVES CAN BE LEFT ON LONG ENOUGH FOR THOSE MINUSCULE PUSHES TO ADD UP TO QUITE A WALLOP.

light. Some experts have therefore suggested using a giant laser to push the sails. This would enable sail-powered spacecraft to reach very high speeds—push a large enough sail with a big enough laser, and you could even reach a nearby star within a human lifetime.

But unlike sunlight, which is free, a giant laser facility would be extraordinarily expensive to build. And there's no guarantee that a sail would survive more than a few years in space, given the danger that interplanetary flotsam could leave the sail in tatters.

A related concept has received some scientific attention of late. Instead of hoisting a gargantuan sheet of aluminum, some physicists have proposed projecting a giant magnetic field around a spacecraft using arcs of electrically charged helium gas. A device the size of a bucket of fried chicken could generate a magnetic bubble 20 miles across. The electrically charged particles in the solar wind would push against this magnetic balloon and propel it and the spacecraft toward the outer solar system.

This concept, known as Mini-Magnetic Plasma Propulsion, or M2P2, is the brainchild of Robert Winglee of the University of Washington in Seattle. Winglee suggests that a small craft using this method could reach speeds of 180,000 miles per hour in three months, without burning propellant or rigging a flimsy sail. Experiments Winglee has conducted show that an M2P2 system could provide a steady thrust while consuming less than two pounds of helium gas a day.

M2P2 propulsion may be tested in space in the near future. But whether they are riding the solar wind or catching a sunbeam, it seems certain that in the coming century, at least a few space-ships will give a new meaning to the phrase "sailing by the stars."

IF WE HAVE the will, then humanity will have the means to conquer the solar system, whether by nuclear rockets or solar sails. But alas, a huge barrier presented by Albert Einstein's theory of relativity remains. According to Einstein, no object can travel faster than the speed of light, and even approaching light speed requires an astronomical expenditure of energy. If our destiny is in the stars, we'll need to find a new kind of physics, or at least a loophole in Einstein's laws.

The physicists are hard at work on the problem. They are pursuing wrinkles in old theories and developing new ones that would enable faster-than-light travel. For instance, one idea that began to receive interest back in the 1970s suggested sending a ship through a hypothetical shortcut between two points in space. First theorized by physicist Kip Thorne, such a shortcut, or wormhole, may form naturally in conjunction with the collapse of a black hole and provide a bridge across the galaxy that doesn't violate Einstein's speed limit.

Another way around Einstein was proposed by Mexican physicist Miguel Alcubierre in 1994. Alcubierre envi-

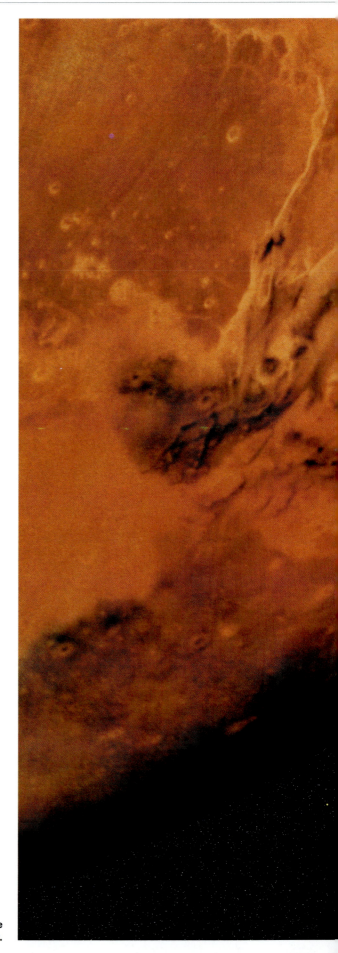

PREMIUM GAS: The ionized, heated gas (plasma) spewed out of the VASIMR can generate speeds of 100,000 miles an hour, making a trip to Mars a relatively quick flight.

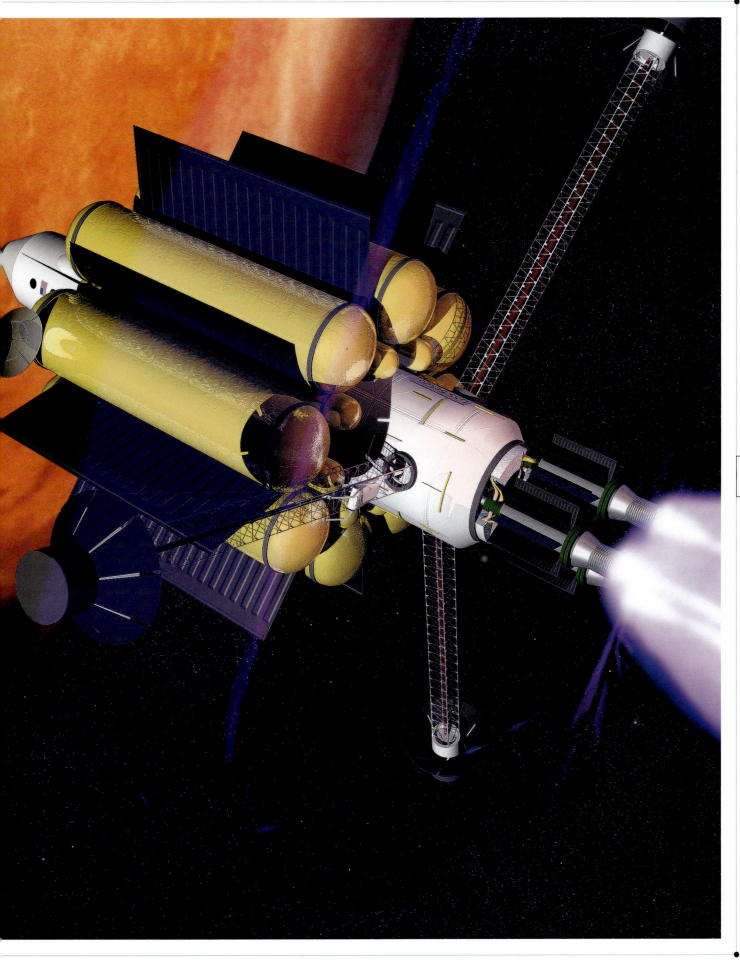

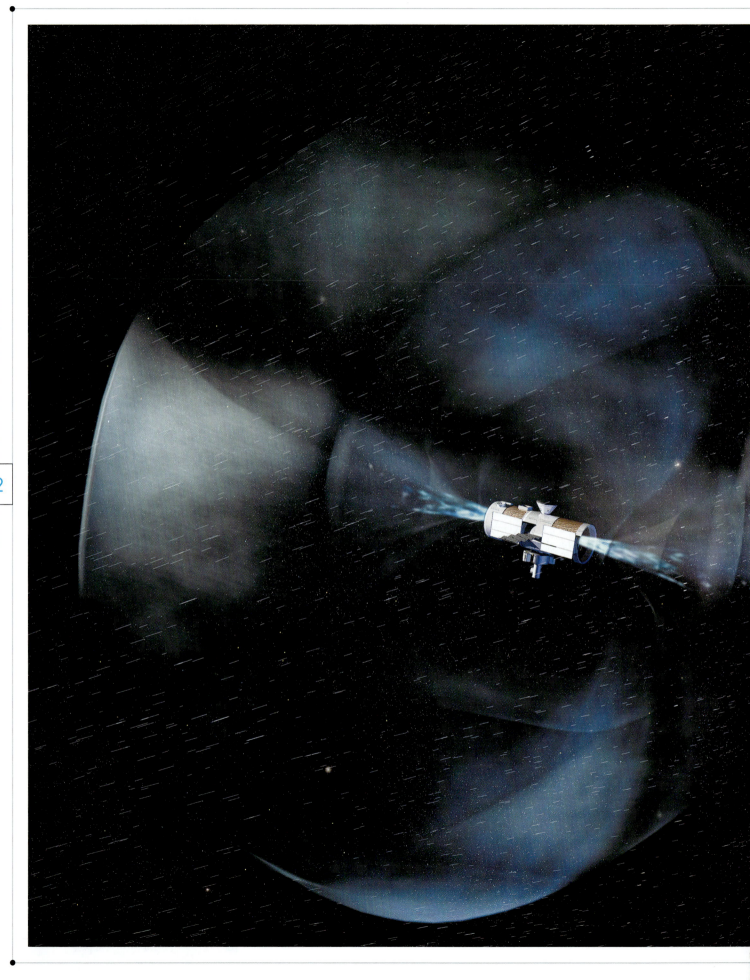

IT'S A BREEZE

No, M2P2 is not R2D2's replacement, but the very efficient Mini-Magnetic Plasma Propulsion system that uses arcs of electrically charged helium gas to generate a magnetic field that the solar wind pushes toward the outer solar system. M2P2 could provide steady thrust using only two pounds of helium a day

sioned a way to warp space-time so that space ahead of a ship shrank away to nothing while space was rapidly being added behind it. Thus, a vessel in a small patch of space could travel across vast distances while—from the Einsteinian perspective—not seeming to move at all. To create such an extreme warp in the fabric of space would require some sort of negative mass, a hypothetical kind of matter that bends space-time in the opposite direction to normal matter. It might be possible, but physicists don't even know where to begin to make such stuff.

Another possibility is to eliminate or even reverse certain attributes of matter, such as inertia—the property that requires a force to create acceleration. In theory, a spacecraft with zero or neg-

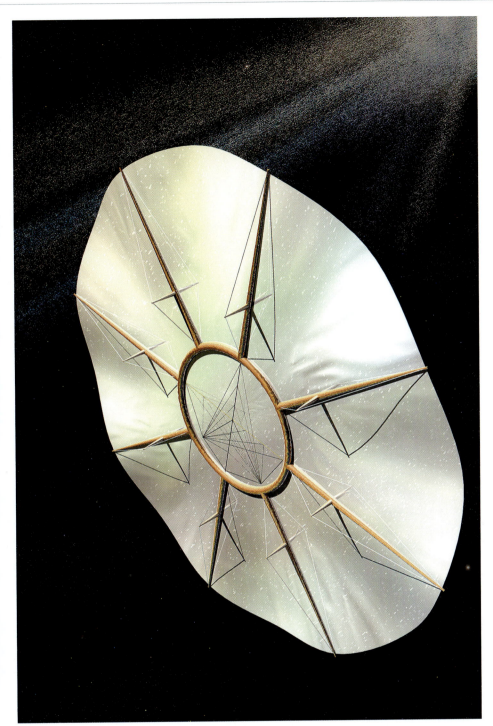

SAILING ALONG: A sail-powered craft (left) might span 440 yards, twice the diameter of Louisiana's Super Dome, so solar sails will need to be light—possibly carbon fiber with a density of 1/10 ounce per square yard, like the material held at far left by Marshall Space Flight Center's Les Johnson—but able to endure harsh conditions.

ative inertia would be able to flout relativity. Physicists are presently testing theories (all of which are controversial) linking inertia to electromagnetism through a phenomenon known as zero-point energy. This energy is created by the constant creation and annihilation of subatomic particles on the smallest scales of quantum space. There are tantalizing signs of progress: physicists in Finland reported being able to create a gravity shield by means of a spinning superconducting disk.

Is faster-than-light space travel really possible? Remember that most experts once believed that heavier-than-air flight was impossible. It took only a few decades to go from the Wright brothers to the Apollo Moon landings. So who can say what could be possible by 2100?

CHAPTER 4

[MAN AND MACHINE]

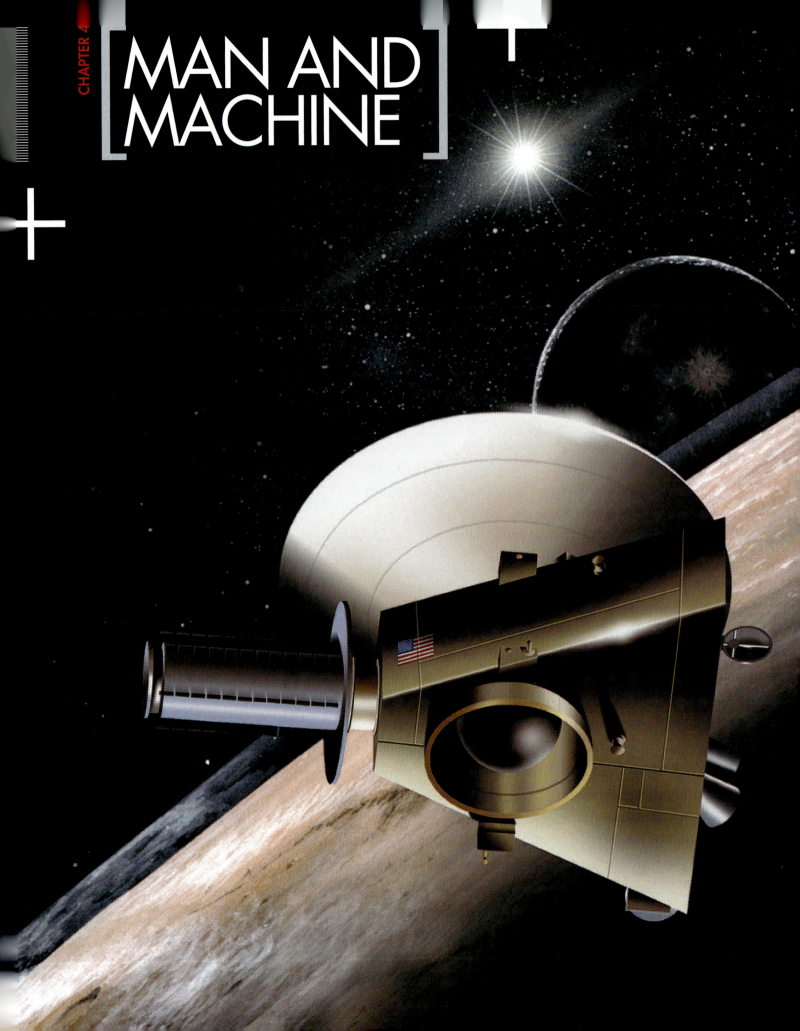

WORLDS AWAY: The New Horizons spacecraft, set to launch in 2006, maneuvers through the Kuiper Belt 12 years later, against a backdrop of Jupiter and Neptune, with the Sun 4.1 billion miles away.

IF WE EVER DO FIND ANOTHER EARTH-LIKE PLANET SOMEWHERE IN THE UNIVERSE, IT'S A SAFE BET THAT THE FIRST EMISSARY WE SEND WILL NOT BE A HUMAN, BUT A ROBOT. EVERY MOVE THAT HUMANKIND HAS MADE IN LEAVING EARTH HAS BEEN ON A PATH PAVED BY A MECHANICAL SCOUT: SPUTNIK 1'S LAUNCH IN 1957 PRECEDED THE FIRST ASTRONAUTS INTO ORBIT, AND THE MOON SURVEYOR CRAFTS IN THE 1960S RELAYED BACK INFORMATION CRUCIAL TO DECIDING WHERE

INITIAL EFFORT: Two days after the launch of Sputnik 1, in 1957, Russian technicians track its orbit on a large globe at the Moscow Planetarium.

the subsequent manned Apollo Moon missions would land. Robots continue to forge the way into outer space, and soon will have visited every one of our neighboring planets. And Voyager 1 may soon be the first machine to leave our solar system.

Why do robots get all the glory? Frankly, because humans aren't up to the tasks yet. Even Mars, probably the most hospitable planet for human exploration, suffers dust storms, 100-mph winds, temperatures that fall to 200°F below zero, powerful radiation, and an atmosphere made of mostly thin carbon dioxide. Traveling in space opens humans to dangers like cosmic ray storms and other radiation. It's no cakewalk for a robot either, but machines have some advantages: Galileo, now orbiting Jupiter and its moons, encountered enough radiation in 2002 to make it put itself into a "safe" mode, shutting down most of its systems in time to save them, although the recording tape on which it stored data nearly melted; commands from Earth got the craft up and running again. Humans don't have a "safe mode" option.

The robotic space fleet is large and varied, from rovers the size of toy cars to orbiters several hundred feet across. Orbiters, which form the bulk of the robots we create for space, are somewhat stand-offish, making their mea-

BETWEEN OUR NINE PLANETS, THERE'S A LOT OF OPEN SPACE, AND NO ONE REALLY KNOWS WHAT'S THERE.

surements without entering a planet's atmosphere, let alone setting down on the surface. But despite the fact that they keep their distance, robotic craft that orbit or fly by planets can use a suite of tools to find out all kinds of detailed information. Cameras, radar, and lasers can do a good job of figuring out a planet's surface features and geology, while other instruments such as spectrometers can decipher the chemical composition of the atmosphere based on how the particles interact with light and other electromagnetic radiation.

Using these sorts of tools, robots have been very successful explorers, relaying unprecedented scientific infor-

mation back to their terrestrial masters. In 1995, when the unmanned spacecraft Galileo fired a robotic probe onto Jupiter, the data it collected upended researchers' beliefs about the composition of the Jovian atmosphere. Galileo itself later made discoveries that have been heralded as some of the most significant of the Space Age, namely its data showing that Jupiter's moons Europa, Callisto, and Ganymede may have vast oceans of water beneath their icy surfaces—conditions that make life possible. This has prompted NASA to plan the Jupiter Icy Moons Orbiter, an unmanned craft scheduled to launch around 2011, to further scan these

SPACE ORBITERS: The Galileo spacecraft (opposite, being launched from Space Shuttle Atlantis in 1989) will end its mission—it has been orbiting Jupiter and its moons since December 1995—on September 21, 2003. Launched on December 11, 1998, the Mars Climate Orbiter (MCO, above) was to coordinate with the Mars Polar Lander (MPL, launched January 3, 1999). The MCO was lost due to a failure to convert from English to metric units in the navigational software, and the MPL was destroyed during its attempted landing.

moons for signs of oceans. The craft would be the first powered by ion propulsion from a small nuclear fission reactor. Heat from the reactor is converted into electricity, which causes charged particles to shoot out of a nozzle at high speed—a propulsion method successfully tested on the Deep Space 1 comet-observing craft in 1999.

Scientists have similar hopes for advances from the Cassini spacecraft, under way since 1997. In 2005, it is scheduled to send a probe, called Huygens, hurtling to Saturn's largest moon, Titan. Shrouded in a thick orange haze that hides its surface, Titan is thought to harbor conditions much like those of a very young Earth. Although its surface temperature hovers at a few hundred degrees below zero, some scientists theorize that Titan may have a liquid ocean made of methane. As it descends into this haze, Huygens's instruments will measure Titan's atmosphere, and a camera and infrared imager will take pictures of the surface once the probe breaks through the cloud deck. Although the probe will have a parachute to slow its decent, it will crash-land into the surface at about 15 mph. If it survives the impact—more likely, if it splashes down into a liquid ocean—the probe will send back data about the surface for approximately 30 minutes, until its batteries die.

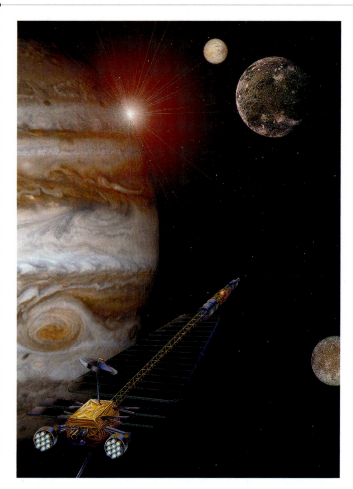

Earth, scientists are eager to see what Pluto is made of.

Between our nine planets, there's a lot of open space, and no one really knows what's there. Soon to be sweeping that territory is Stardust, launched in 1999 to collect interstellar dust and also material from a comet's tail. The spacecraft is collecting the dust on three orbits around the Sun and will fly through Comet Wild-2 in 2004. The spacecraft will then return to Earth with its microscopic cargo in 2006. Although space dust is smaller than grains of sand, it is moving very fast, so Stardust has to capture it without heating it as it slows down, as this will alter the dust's chemical composition or vaporize it entirely. The answer is aerogel, a transparent, sponge-like substance made of puffed silicon. This is fitted into a collector, shaped much like a tennis racquet, which unfolds from Stardust to expose it to space. When dust hits the collector, the aerogel dissipates the energy of impact much like a car bumper in a traffic accident, bringing the dust to a gradual stop. When the spacecraft comes back to Earth in 2006, it will parachute down in a reentry capsule with the collector aboard. This will be the first time that extraterrestrial material has been returned to Earth from outside the orbit of the Moon.

Over the next few decades, there are plans to send robotic craft outside the boundaries of our solar system. The solar wind creates a bubble around our system, called a heliosphere, which shields us from interstellar magnetic fields, cosmic rays, and dust. No one is really sure what size the heliosphere is, but it's theorized to be at a distance 80 to 150 times the distance from the Earth to the Sun. A craft called simply Interstellar Probe is planned to interrogate that bubble to see how stars like ours interact with their environment, and then press on outside of the heliosphere to see what the uni-

To prepare the probe for its mission during its six-year transit, NASA engineers send it a wake-up call every six months and have it exercise its moving parts and perform a systems check of its electronics and programming. So far, everything seems ready to go.

The last of our nine planets, Pluto, is the only one yet to receive a visit from an Earthling spacecraft. NASA hopes to rectify this with New Horizons, a mission set to launch in 2006 that will take some 12 years to reach Pluto and its moon, Charon. Unlike the other planets in our solar system, Pluto is made of the same matter as Kuiper Belt Objects, the cloud of small, icy bodies orbiting past Neptune that are made of material left over from when the other planets formed. Since the planet consists of this primordial matter that set the stage for the evolution of the solar system, including life on

MANY MOONS: The Jupiter Icy Moons Orbiter, an ambitious mission proposed for 2011, would orbit three planet-size moons of Jupiter—Callisto, Ganymede, and Europa—which may conceal vast oceans beneath their icy surfaces. In 1996, the Galileo spacecraft found evidence on the moons of water, energy, and the necessary chemical ingredients essential to sustain life.

TAKE A MEMO

The personal assistants of the future will be on the job 24/7

SCIENCE FICTION HAS LONG teased us with the prospect of our own personal robot, one as loyal as a dog, only smarter, able to change the oil in a car, vacuum the living room, or scramble up a steep-pitched roof to clean the rain gutters. In short, to do all those tasks that humans find too dirty, dull, or dangerous.

NASA wants the same thing for its astronauts. They're not looking to replace humans; indeed, in its 2003 Strategic Plan, the agency flatly states: "The capacity for complex reasoning, intuition, and learned observational skills will not be available in robots for many decades, if ever." Instead, NASA wants robots to work in concert with humans, serving as their assistants.

As computers and software continue to get more powerful, more and more prototypical "smart" and somewhat autonomous robots are being developed worldwide that may do just that. Two in particular stand out, if for no other reason than they look like something that stepped off a movie screen.

One is the Personal Satellite Assistant (PSA), and it was indeed inspired by a movie. The PSA is a softball-size sphere that floats, and is designed to hover near a working astronaut or near in the microgravity of the Space Shuttle or the International Space Station (ISS). It is being developed at NASA's Ames Research Center, in California, and was the brainchild of NASA scientist Yuri Gawdiak, who admits the idea came to him after watching the scene with the floating orb that honed Luke Skywalker's light-saber skills in the original Star Wars.

The PSA will use fans and blowing air to move along independently, at a speed of about a yard a second, through a spaceship's interior. It's equipped with stereo cameras and a video display, a microphone and speaker, and various sensors, all contained in a hard plastic cover (although the final version will have an additional soft cover around the plastic). It will be able to respond to voice commands, and do the routine but critical task of monitoring the environment, checking such things as temperature, humidity, and oxygen level, as well as looking for leaks.

FOLLOW THE BOUNCING BALL: Personal Satellite Assistants, or PSAs, may soon join PDAs as essential to the technologically current life. But instead of riding in your pocket like PDAs, PSAs will ride alongside you, on their own power, recording data and responding to voice commands.

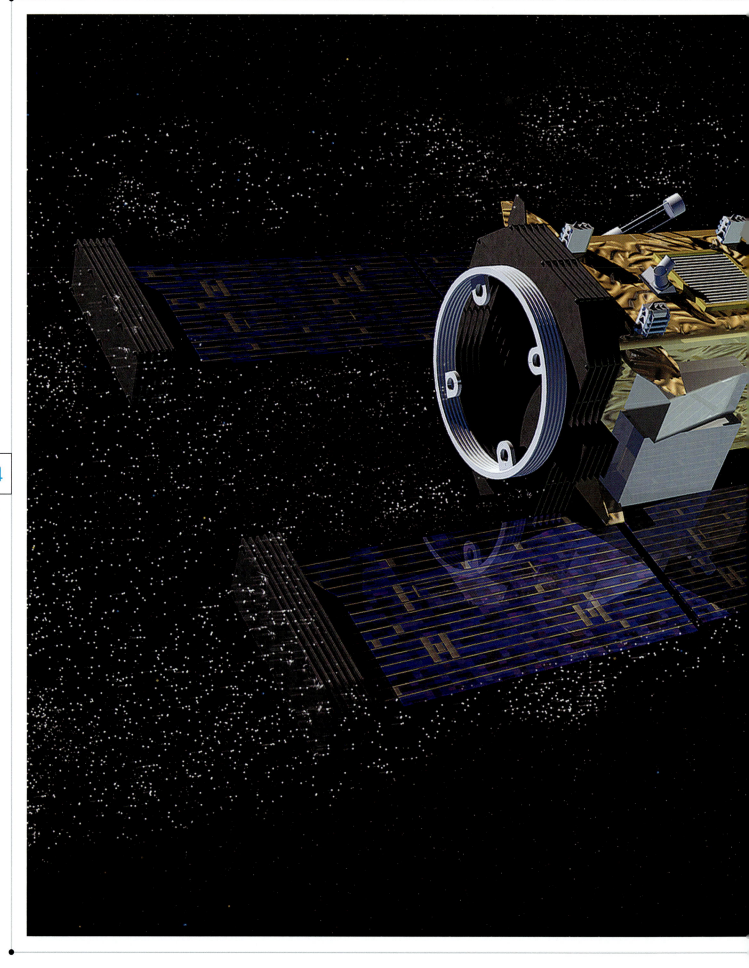

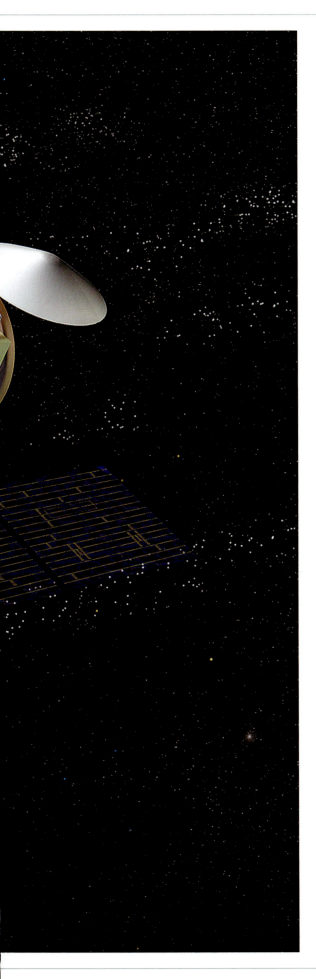

DUSTBUSTER: Stardust (left) was launched in 1999 to collect interstellar dust as it orbits the Sun. On its scheduled return to Earth in 2006, scientists will get a look at its minute and dusty cargo. In order to capture the dust safely, without its molecular structure being changed by the heat generated during deceleration, the dust collector is filled with aerogel (above), which is made of puffed silicon.

verse really is like on the other side. Once it has cleared the interference of the heliosphere, scientists hope that the craft can also get a better look at the cosmic infrared background radiation, energy left over from the Big Bang that tells us about the origins of the universe.

As we send our robotic explorers farther and farther out into the universe, one plan is to stop sending one big craft, but instead release a swarm of smaller vehicles that work together like ants in a colony. These mini-probes can fan out to cover a much larger area than one large craft could see on its own, and if a few of them were lost to the hazards of space, the rest could compensate and finish the mission. Smaller probes are also easier and cheaper to build, so they would cost less to launch. One proposed micro-fleet is the Autonomous Nano Technology Swarm, or ANTS, which would be sent to prospect the asteroid belt that lies between Mars and Jupiter, looking for resources that could aid a space colony. ANTS would consist of about 1,000 probes, each weighing about two pounds. Groups of 100 ANTS would carry the same surveying tool, such as a magnetometer or radar, and they would

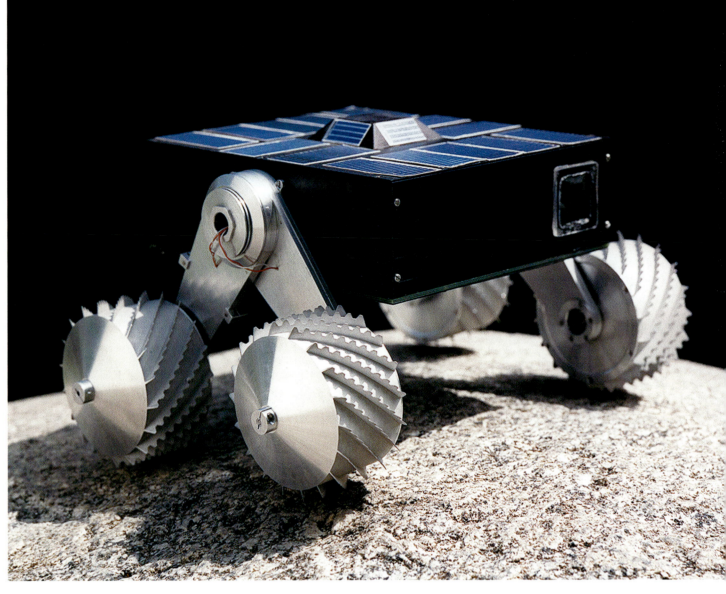

ROIDS RAGE:
Compact nano rovers (above) could transmit data, including pictures with panoramic to microscopic views, from an asteroid like the one opposite. An asteroid, defined as a lump of rock or a small planet that follows an elliptical orbit around the Sun, may have many other asteroids around it.

fan out among the space rocks in order to make a detailed survey of the metals and other resources there. The worker ANTS would radio their data back to a set of 100 ruler and messenger ANTS, which form the game plan that guides the prospecting mission. When the survey is complete, only the few messenger ANTS would need to make the return trip with the acquired data.

The engineers who dreamed up the ANTS mission envision it launching around 2020, but they are assuming that by then, they will be able to launch the ANTS fleet from some kind of

human-occupied space station at some distance from Earth. If that is going to be the case at any point, let alone by 2020, we'll probably need to send more robots to visit the surfaces of other planets.

If a colony were to be set up on, say, Mars, most scientists agree that the best chance of success would be to have a bunch of robots land several years before and prepare the site for humans, by building shelters and gathering fuel. But the reason why we've had more orbiting robots than surface-exploring ones is that landing is probably the

most dangerous task for a robot. Conditions on the surface are never going to be exactly as anticipated, and machines running pre-programmed landing sequences don't have the ability to adapt to unexpected situations. In 1999, the Mars Polar Lander was lost during its attempt to land on the Red Planet, and researchers still aren't entirely sure what went wrong. But like all of space travel, it's a risk we're going to have to take, and learn how to handle better, if we're going to continue to advance exploration.

In the summer of 2003, NASA launched the latest pair of Mars rovers, Spirit and Opportunity, which look much like overgrown cousins of the 1997 Sojourner rover. Future Mars rovers may be more specialized. There are plans for ones that look like mini-bulldozers, with a scooping arm and a hopper on top of the robot. Other rovers

THE REASON WHY WE'VE HAD MORE ORBITING ROBOTS IS THAT LANDING IS THE MOST DANGEROUS TASK FOR A ROBOT.

can come equipped with arms that can grasp and hoist objects; two of these could cooperatively lift construction supplies, such as long metal support beams. The bulldozer rovers could dig out a hole for a sub-surface habitat, then move on to an unexplored part of the

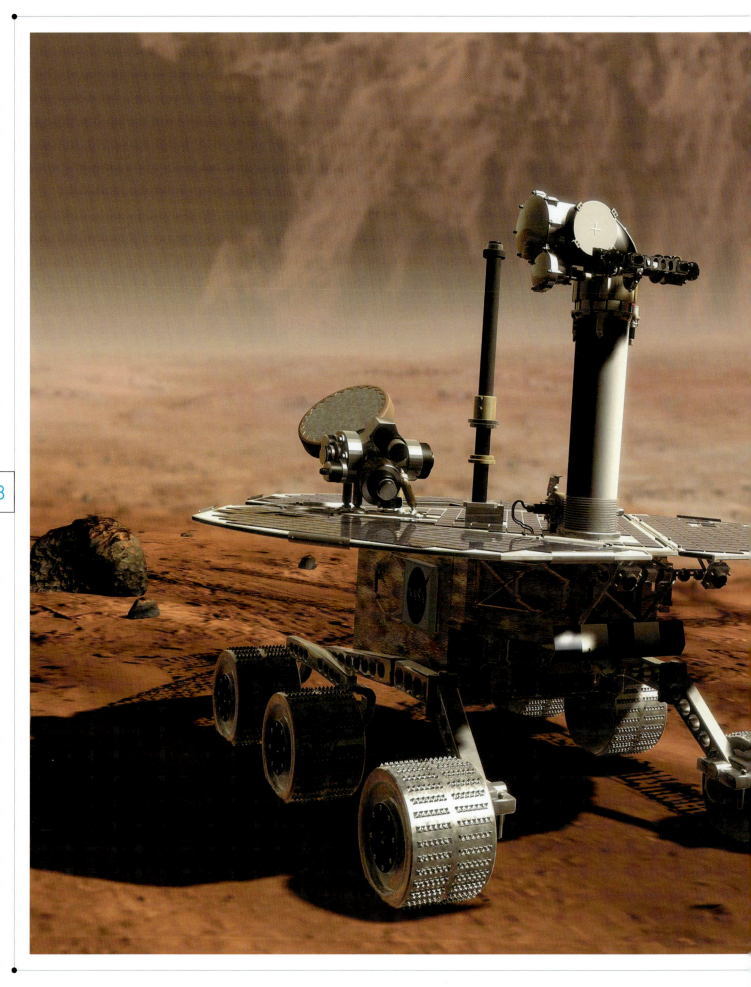

BULLDOZER ROVERS COULD DIG OUT A SUBSURFACE HABITAT, THEN MOVE ON TO AN UNEXPLORED PART OF THE PLANET.

planet for an archaeological dig, while the work crew robots finish construction of the shelter. Some distance away, at the Martian poles, a Cryobot could be digging for water. The three-foot-long cylindrical droid has a heated nose to melt through ice, and then falls ever downward under the pull of gravity. Ice would refreeze behind the Cryobot as it descends, which lessens the chance of contaminating the exploration site with material from the surface. In tests in the Arctic, the prototype Cryobot melted through 75 feet of glacier, and an onboard camera and chemical sensor allowed the probe to analyze the subsurface layers without having to haul a core to the surface.

Other rovers, such as a team of cliff-climbing robots, could be more specialized for exploration tasks. Two anchor robots at the top of a sheer face play out tether lines as the third cliff-bot rappels over the edge, all sharing information from their on-board cameras and sensors to guide the descent. The robot climbing team can get into deep gullies that would

RED ROVER, RED ROVER: Mars Exploration Program rovers Spirit (launched June 10, 2003) and Opportunity (launched July 7, 2003) are heading for opposite sides of Mars. These rovers are larger cousins to the 1997 Mars Pathfinder rover.

LET THE ROBOT DO IT

The talented Robonaut will function as the eyes and ears of future explorers

LIKE THE PSA, ROBONAUT ALSO looks like a *Star Wars* character. Under development at NASA's Johnson Space Center, the robot has a vague resemblance to a human, but one with a gold-colored head, two arms, a distinctly V-shaped body and a single leg down its center.

While the PSA is currently intended for inside use, teams of Robonauts will eventually work outside the ISS or a spacecraft, in collaboration with astronauts. While such "extravehicular activities"— space walks—are fascinating for those of us on Earth to watch, they're also dangerous. An astronaut can become snagged on something, slip, or be hit by a micrometeorite. Better to let the robot do it.

Robonaut is fitted with the expected accoutrements— numerous sensors, stereo vision for eyes—but what are unique are his lifelike hands. The robot has four fingers and a thumb, with 14 degrees of freedom in each hand. That means his hands can twist, tighten, bend, and grasp with the best of them (human hands have 22 degrees of freedom). That allows it to use tools intended for human hands,

from tweezers to a power drill, which means Robonaut could eventually work outside doing repairs or routine maintenance, or eventually even assembling orbital telescopes or Earth observatories.

What Robonaut lacks is autonomy. For now, the robot would be controlled through "telepresence," in which an astronaut safely inside a spacecraft would wear virtual-reality glasses and move the robot's arms and fingers using gloves wired with sensors. The robot would serve as the astronaut's eyes, and mimic the operator's motions.

No word yet, though, on how well it handles a light-saber.

R2-D2 MEETS C-3PO: An astronaut, looking like R2-D2 of *Star Wars* fame, shakes hands with Robonaut, a very human-looking version of a robot that looks in turn like C-3PO. While robots designed for specific tasks can look odd and different, we will want those that spend time with us to look more familiar and less techy.

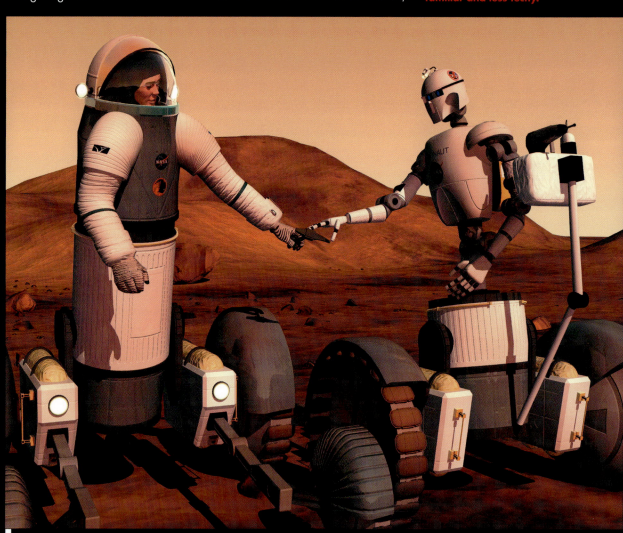

CLIFFHANGERS: Of the various robotic prototypes, the "cliffbot" is designed to handle the steepest terrain. These robots would work alone or in teams of three. With two serving as anchors on the cliff edge and playing out tethers, a third would be able to maneuver the almost-vertical cliffs of Mars, in search of rich water-borne mineral deposits.

be the end of other rovers, and they can also examine the stratification of cliff walls to understand more about the geological history of the planet.

Another proposed way to look at geological formations and cover large distances quickly is with robotic microflyers. Some researchers envision miniature blimps, a few feet in length, that would glide about the Martian skies and relay back images of potentially interesting locations. Other prototypes under development include a triangle-shaped plane with a 34-inch wingspan, which uses an artificial vision system inspired by dragonflies and bees. Eight photodiodes at the front of the craft detect changes in light levels and pick up the horizon line in order to keep the craft on a level flight, mimicking the function of a dragon-

fly's simple eyes. Meanwhile, three cameras send data to a microprocessor that measures speed and distance the way bees do: Objects that are close appear to move more quickly than those that are far away, so the system can translate speed into distance. In Earthbound trials, the flyer has successfully used these systems to control its own flight.

Such biomorphic robots—those with designs inspired by the principles of biologic systems—have the potential to mimic all kinds of animals and plants, from the fluttering glide of a maple seed to the slithering of a snake. One design for a hopping rover takes its cues from amphibians: The rover has froglike metallic legs and a spring attached between the "knees," which pulls taught when the rover bends its legs; when

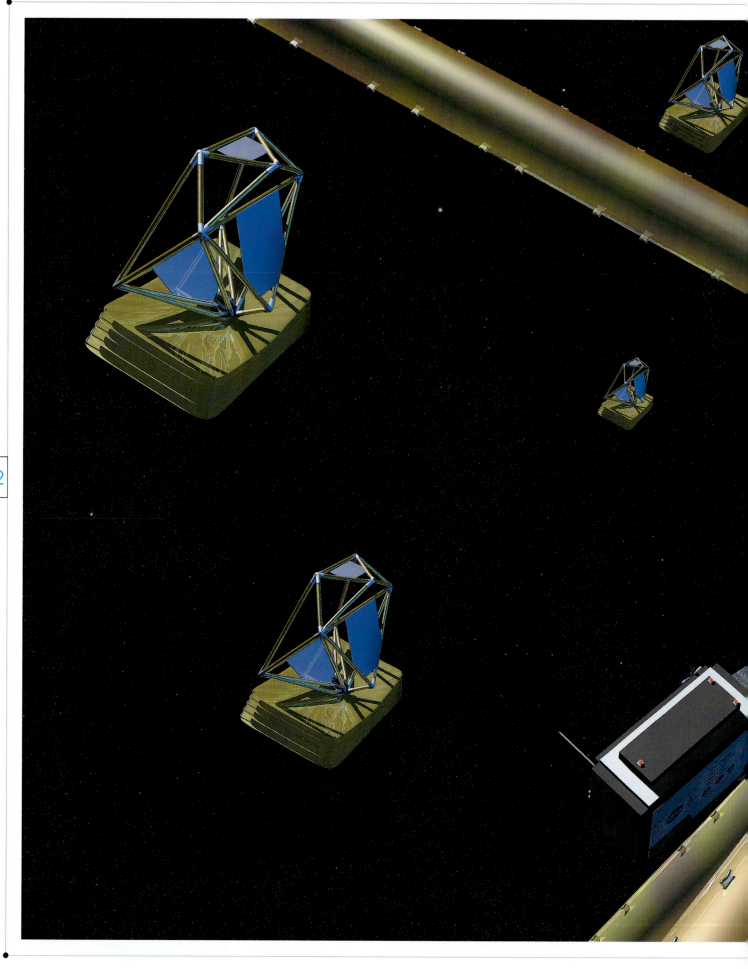

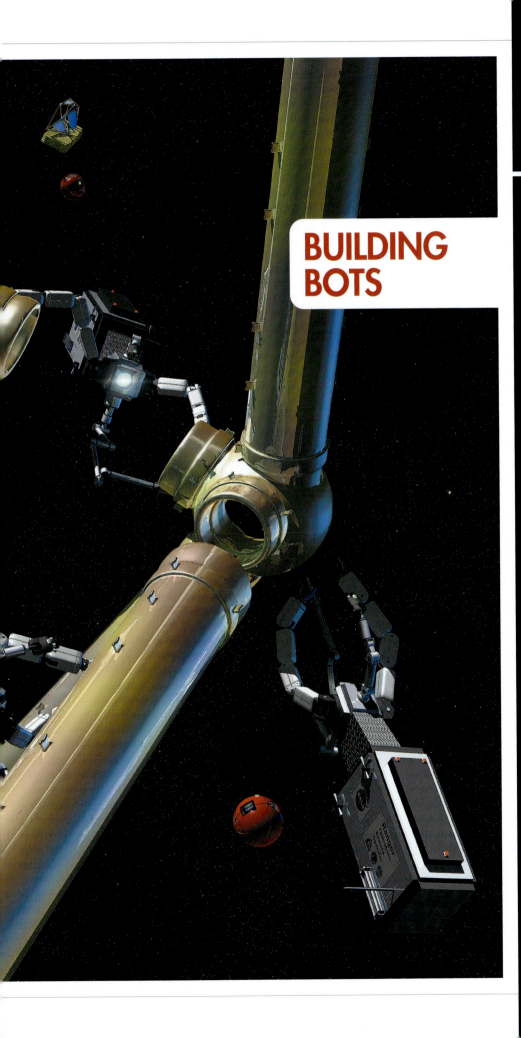

BUILDING BOTS

Autonomous construction robots inflate and connect the large and small pieces necessary to assemble a telescope. During the assembly process, free-flying robotic satellite assistants, larger versions of PSAs, function as inspection robots, able to move around and see from different angles. Floating in the background are the already assembled telescopes, forming an array.

SPACE CRITTERS: Low-cost, small, and lightweight, biomorphic explorers (right) borrow designs from nature and fly like bees, flutter like maple seeds, or burrow like worms. Hopping rovers (opposite) are small, three-pound, frog-like robots designed to hop up to twenty feet by use of a spring between their "knees." Despite its size, each will carry sensors, a camera, solar panels, and a computer.

released, the robot could jump 20 feet in Mars's reduced gravity. Another spherical, inflated rover, called Tumbleweed, looks much like an oversized beach ball. The robot would inflate itself as it entered the Martian atmosphere, acting as its own airbag upon impact with the

a halt. The rover might also work well on Titan, since its buoyant nature would keep it afloat on methane seas.

Since there are so many different conditions on planetary surfaces, the best rover might be one that could morph to fit the domain. One prototype is Polybot,

surface. Inside Tumbleweed, secured by tension cords to the sphere, would be a payload of scientific instruments. Winds would cause the inflated rover to bounce over the surface, measuring its surroundings as it went. If it were to see something interesting, say, if its radar were to pick up subsurface water, it could partially deflate to bring itself to

made from multiple hourglass-shaped modules, each with a hinge in the middle and equipped with its own microprocessor and motor. The components can link up to form a speedy tank-tread loop, then shape-shift into a snake for tight places, or a multilegged form to clamber over obstacles. Researchers in Switzerland have taken this multi-tasking to the next

level, with a swarm of four-inch-wide rovers called s-bots. Each s-bot has two gripper arms and a host of sensors for monitoring the environment. The s-bots can fan out over an area and communicate with each other via radio or by a ring of lights around their outsides. Two s-bots can link arms to hoist each other over gaps, or the whole squad can link up to cooperatively move an object.

With all these grandiose plans, there are some critics who argue that robots are not advancing quickly enough to fulfill these goals. For example, the average laptop computer has about 1,000 times more computing power than the two Mars landers launched in 2003. And power is also an issue, says Charles Elachi, director of the Jet Propulsion Laboratory in Pasadena, California, NASA's lead center for robotic exploration of the solar system. "We spent $800 million on those two rovers and they're going to work for three months," he says. "If we had a nuclear-powered system instead of solar, they would have worked for five years."

Rovers have had conservative abilities for good reason. Older chips have been tested and are more reliable, an essential property for space travel: If a computer crashes when it's several hundred million miles from Earth, there's no one to hit the reboot button. With gaffes like the 1999 crash of the $125-million Mars Climate Orbiter, which presumably burned up in the Martian atmosphere when scientists neglected to convert critical navigation data from English to metric units, NASA wants to avoid any further embarrassing mistakes. Despite this, Elachi believes that as early as the next decade or two, we could have one or more robotic outposts that work in a similar way to the research station the National Science Foundation maintains in Antarctica.

THE AVERAGE LAPTOP COMPUTER HAS ABOUT 1,000 TIMES MORE COMPUTING POWER THAN THE TWO MARS LANDERS.

Robots will have to get far more intelligent for this to happen. While a computer can be programmed to win a game of chess, it's still very difficult to imbue a robot with enough smarts to recognize and avoid obstacles like rocks, or tell the difference between a shadow and a hole in the ground. But Richard Terrile, a JPL scientist, believes we will

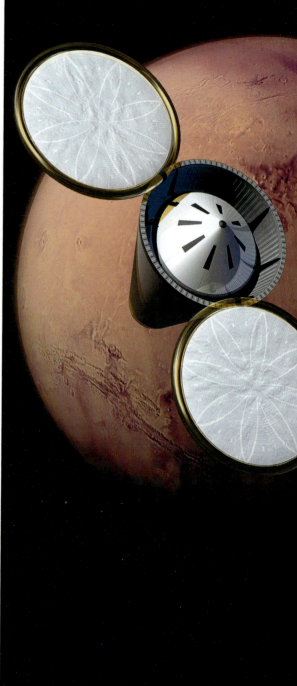

have fully autonomous, smart robots sometime this century. Terrile heads the Mars Scout Mission, a series of small science missions to Mars that are planned to start launching in 2007. Terrile's group uses what's called "evolutionary computation" to develop intelligence for space robots. The idea is to let loose a bunch of small computer programs, each of which tries a different way to solve a problem. Those that form some kind of answer are the "parents" of the next generation of programs, recombined randomly to see if their offspring do any better. After many generations, the programs become very effi-

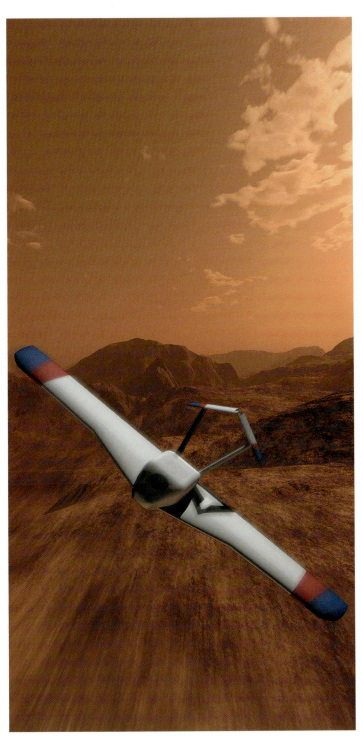

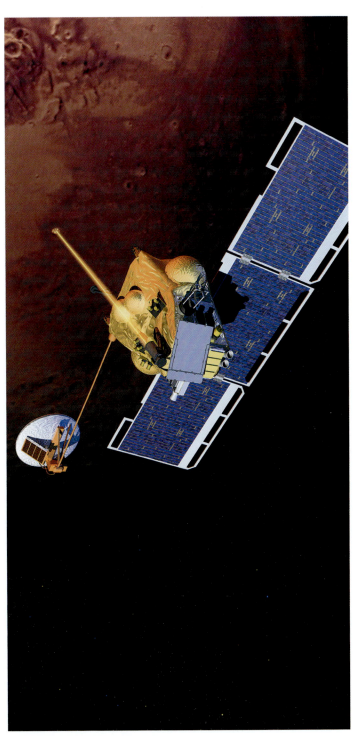

cient at solving the problem. Some NASA engineers envision using this technique to get around a rover or spacecraft's lack of adaptability; the machines are stuck with whatever software they had when they were launched. But with evolutionary software, the probes could adopt new programs, refine them under the circumstances they encounter in space, then beam the improved version back to Earth for the next mission. If this becomes routine, by the time we do send a robotic emissary to a far-flung Earth-like planet, it might have enough intelligence to make a good ambassador.

A SHUTTLE SHIFT: Three teams have been funded to develop an Orbital Space Plane (OSP), the next-generation reusable-launch system intended to replace the Space Shuttle.

THE SHAPE OF THINGS TO COME

THE SPACECRAFT OF THE FUTURE WILL BE DESIGNED FOR ONE OF TWO PURPOSES: SENDING THE REGULAR SCHMO ABOVE THE EARTH OR SENDING AT LEAST ONE INTENSELY TRAINED SCHMO A WHOLE LOT FARTHER. THE ONLY WAY TO GET THE MASSES EXCITED ABOUT A SIMPLE ORBIT THESE DAYS IS TO PUT THEM IN ONE. (EVERYONE KNOWS DENNIS TITO'S NAME. HOW MANY PEOPLE KNOW THE NAMES OF THE ASTRONAUTS IN THE ISS?) TO DO THAT, FLIGHTS TO SPACE HAVE TO BECOME MUCH, MUCH CHEAPER.

HIGH COST OF TRAVEL: For the average Joe who can't wait and has a few extra dollars, Space Adventures offers a suborbital flight for $98,000.

That doesn't mean we don't need well-funded astronauts pioneering space to make ourselves feel lofty—it's just that they'll have to go farther than low-Earth orbit, and farther than the Moon, to get us excited. And getting us excited will require undreamed of amounts of cash. So the future of how we build spacecraft is a bifurcated one: While governments and giant corporations will spend billions trying to find ways to keep astronauts alive in space for years at a time, backyard rocket enthusiasts will toil to put their friends above the atmosphere.

CHEAP PEEKS

In 1908, the Ford Motor Company wowed the world when it assembled the first Model T, and the average Joe, in addition to being wowed, wanted one. The average Joe got one. In 19 years of production, Ford put out more than 15 million Model Ts.

In 1962, the average Joe was wowed again, when NASA put John Glenn into orbit. And again the average Joe wanted in. But it's been more than 40 years since that flight, and Joe has yet to even fly above the atmosphere. Joe is frustrated.

The X Prize contestants (see page 73), of course, are trying to remedy the situation. So far, their blueprints are certainly innovative. In terms of getting up, the teams all use a rocket of some sort, but when it comes to getting down, the designs get funky. The overall shape of a spacecraft is largely determined by its propulsion system, but thanks to the challenge of reentry, X Prize entrants will make outlandish landings. Possibly ahead of the others in development is Burt Rutan's company, Scaled Composites. Its SpaceShipOne pops above the atmosphere, like a prairie dog having a look around, for a mere three and a half minutes. The cabin is pressurized, so the passengers, unencumbered by space suits, can comfortably peer at the black sky or the planet below them. Shortly after the craft crosses the apogee of its trajectory, the wings bend in half, giving it some measure of stability. Unlike the Space Shuttle, which has to enter the atmosphere at a precise angle (using a handful of expensive, interwoven systems) or risk being tossed about at the mercy of the wind, SpaceShipOne's cocked tail keeps it—dandelion-like—in one position for the whole glide home. Scaled Composites is also planning to use a material never before used to slow a spacecraft—wood. Just before touching down, a maple skid will emerge from beneath the cockpit, then scrape across the runway, bringing the ship to a halt.

SpaceShipOne's chief competitor, the Black Armadillo, looks like a pencil on its last half inch. It's the product of John Carmack, the programmer who created the computer game *Doom*, and its method for getting to space is conventional enough—where you'd expect to find an eraser there are four gyroscope-

SCALED COMPOSITES IS ALSO PLANNING TO USE A MATERIAL NEVER BEFORE USED TO SLOW A SPACE-CRAFT—WOOD.

controlled engines ready to blast the thing off the ground. And the initial descent, point downward with the passengers facing the Earth (for some reason the people at Armadillo Research are more worried about G forces on the way back than during the launch), uses a regular old parachute. But the final thud is softened, if all goes well, when the tip collapses upon itself at impact.

Then there's TGV Rocket's Michelle-B—a hexagonal column topped by an R2-D2 look-alike. At re-entry, the skin of the column, what they're calling an aeroshield, opens like an upside down umbrella. A pilot can steer the shield until the last moment, when the engines, in reverse mode, will cushion contact.

Other designs run the gamut from ordinary planes and '50s-style comic-book rockets to UFO-like disks and lightweight apparatus hauled most of the way by balloons.

But private innovators and visionary renegades aren't the only ones trying to make the schlep to space cheaper and more frequent. NASA has spent billions in the hopes of finding snazzy new

MICHELLE, MA BELLE: Modular Incremental Compact High Energy Low-cost Launch Experiment, or Michelle-B, is TGV Rockets' X Prize entry. A vertical takeoff and landing (VTOL) craft, Michelle-B will slow down using a flexible aeroshield that will allow the pilot to steer.

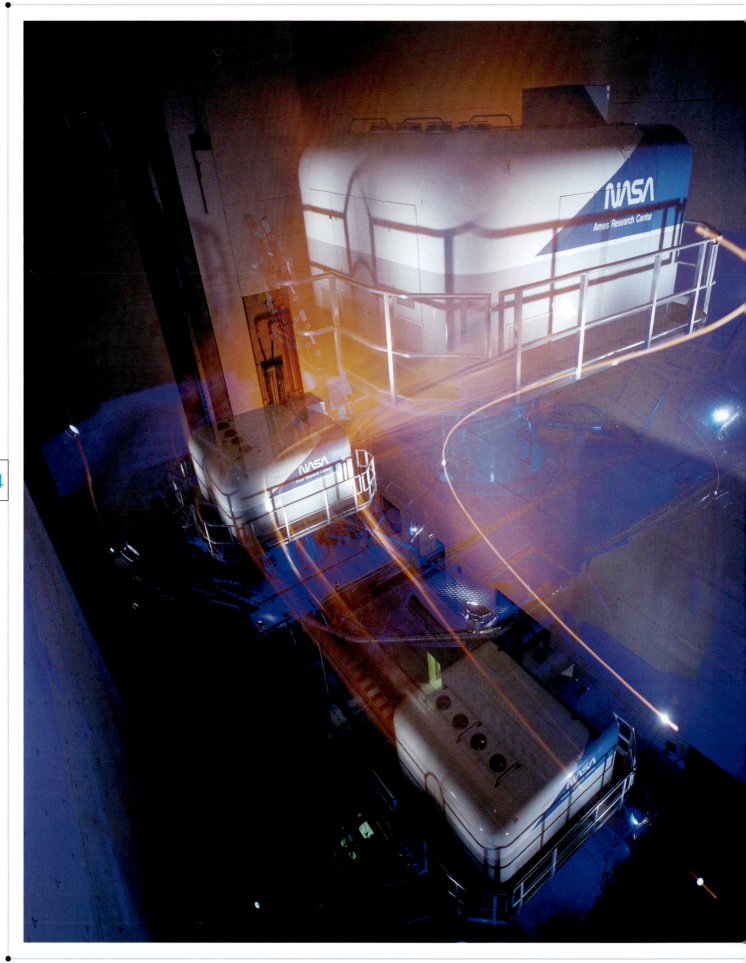

ways to get above the atmosphere. The X-33 and the Venture Star, both unmanned, looked something like slightly squashed sno-cones with little wings stuck on in the back. But plans to launch the vehicles were canceled when the programs went over budget.

In 2001, NASA started the Space Launch Initiative (SLI) program to find a replacement for the Space Shuttle. The Shuttle was originally supposed to fly 30 to 50 times a year and make a bit

gle human monitor at ground control, the new designs had updated, inter-communicating networks. The Shuttle's tiles were to be replaced with durable nickel-alloy shingles. And the computers on these ships could jettison the crew in an ejectable cockpit, should they sense a life-threatening failure on some part of the ship.

Unfortunately, NASA scrapped the program in November 2002 due to budget cuts. After the disintegration of the

NASA'S FUTURE PLANS FOR GETTING TO THE INTERNATIONAL SPACE STATION ARE UP IN THE AIR.

MOTION SLICKNESS:
A timelapse photograph shows the Vertical Motion Simulator (VMS, left), which is endowed with the greatest motion range of any flight simulator in the world, making a dramatic move. The cockpit (above) shows the virtual scene from the pilot's seat. Engineers can program the VMS to accurately simulate any aerospace vehicle, thereby allowing pilots to experience realistic cues and aircraft responses.

of money by retrieving and repairing damaged satellites. But commercial interest wasn't what NASA had anticipated, and the Shuttle launches went up only four or five times a year at a cost of $500 million a pop. The heavy weight, the fragile tiles, and the lack of a viable emergency exit strategy for the crew compounded the problems. Under SLI, Lockheed Martin, Northrop Grumman, and Boeing all came up with sleek new concepts for getting a reusable vehicle into orbit and back. Some had the manned craft riding piggyback on a massive jet, and others had it on the tip of long rockets. Where the Space Shuttle had a slew of separate electronic systems, each needing a sin-

Shuttle Columbia in February 2003, this looks to have been a poor decision, and as of this writing, NASA's future plans for getting to the International Space Station are up in the air.

Throwing away a billion dollars to test a single vehicle that may never actually be used is perhaps an inefficient way of coming up with the right spacecraft to replace the Shuttle. In December 2002, NASA came out with a new design paradigm. The Virtual Flight Rapid Integration Test Environment (VFRITE) lets vehicle designers alter virtual spacecraft while a pilot flies their creations in a flight simulator, called the Vertical Motion Simulator (VMS). The VMS can move 40 feet side to side and

STRONGER THAN STEEL

Nanotube technology may allow spacecraft shells to repair themselves in an instant and avoid future catastrophes

WHETHER OR NOT CONCRETE finds a place in otherworldly habitats, there is undoubtedly a need for super light, super strong materials. This need may be filled by the carbon nanotube. The thickness of a nanotube's wall is that of a single atom, giving it incredible strength, stiffness, and lightness. Eventually, entire spacecraft may be made

from this stuff. Jonathan Dordick, a chemical engineer at Rensselaer Polytechnic Institute, may even have found a way to make nanotubes self-healing by stuffing them with light-weight proteins. Wherever one of these nanotubes might break, the protein is released, making an instant adhesive which rejoins the breach, significantly reducing the

risk from space debris. There's a one percent chance that some orbiting chunk smaller than a centimeter will pierce the International Space Station during its twenty-year existence. Should that happen, the station's air would be instantly sucked out through the hole, turning any astronauts inside into jelly. But Dordick's protein could heal the hole behind the piece of debris as it passed through the nanotube wall, keeping inside and outside separate, despite the projectile's initial speed of nearly 16,000 miles per hour.

NANO NANO: A computer graphic (below, left) shows a molecular tube, an example of nanotechnology, which involves the construction of devices at the molecular level. Each colored sphere represents a single atom—carbon (blue), oxygen (red), and hydrogen (yellow). A colored transmission electron micrograph (bottom) shows the capped end and multi-layered wall of a nanotube. Nanotubes consist of sheets of carbon atoms arranged in hexagons (below).

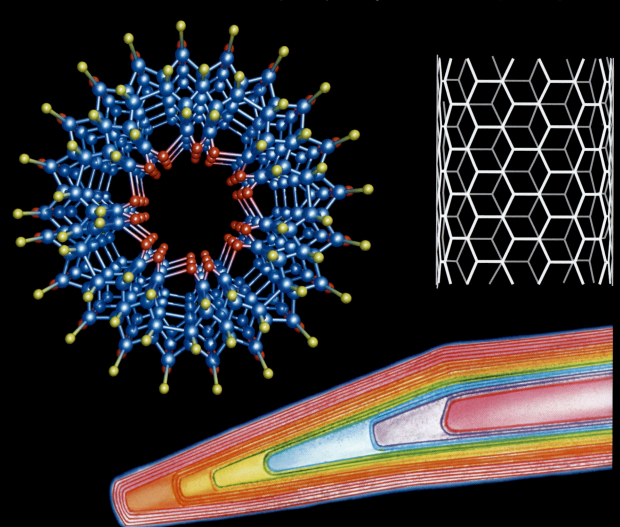

HOLE IN THE WALL:
NASA investigator Dan Bell examines a 16-inch hole in a carbon-reinforced wing panel removed from Space Shuttle Atlantis. The hole was made by a 1.67-pound piece of fuel-tank foam insulation shot out of a 35-foot nitrogen-pressurized gun during a test in San Antonio on July 7, 2003. The test was part of the investigation into the cause of the Shuttle Columbia disaster on February 1, 2003.

60 feet up and down—plenty of room to handle whatever contraption a designer might come up with. With the pilot reporting in real time how the new dream machine handles in the virtual wind, the designers can tweak surfaces, shapes, materials, or sizes—on the fly.

MATERIAL WORLDS

The Space Shuttle Columbia appears to have broken up because a piece of foam hit the wing during lift off. The wing was made of Reinforced Carbon-Carbon (RCC) panels allegedly able to withstand such an impact. And no one expected that the rubber foam would fly off the way it did. "The space community may think they know how materials behave in space, but they

don't," says David Nixon, founder and CEO of Astrocourier, a company that sends experiments into orbit aboard the Space Shuttle. As part of the post-mortem, NASA fired a piece of foam at a simulated wing—and put a hole in it. Further investigation led to the discovery of huge, unanticipated air pockets under the foam. The implications extend beyond the specific materials on the Shuttle to all the lightweight alloys and advanced materials of future spacecraft. "The lesson is that the space community has to test these materials to oblivion before they use them," says Nixon.

To remedy the lack of data, Nixon and his company will send up a host of materials in 2006 to see how they

FOR A ONE-YEAR JAUNT IN SPACE, A SINGLE ASTRONAUT NEEDS ABOUT 10.6 TONS OF WATER.

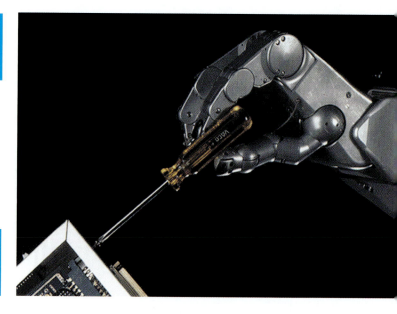

HAND ME THAT SCREWDRIVER: The robotic hand at work (above) demonstrates the precision and power combined in an integrated system. A robot could be programmed and "taught" to pick up and use tools much like its human counterpart.

behave in orbit. Not only will various polymers, alloys, and "aramid fiber meshes for lunar regolith shielding containment" have their day in space, but other neglected, more basic construction materials will be tested as well. Hardwoods, softwoods, rocks, and concrete, up to now unstudied, will be subjected to G forces, vibration, zero gravity, temperature extremes, radiation, and all the other hazards that come with a trip to space. Future space stations may be made of molded graphite-epoxy composites, or they may be made of teak.

STAYIN' ALIVE

A spaceship traveling from the planet's surface to low-Earth orbit has to be able to withstand the stresses of intense acceleration and the friction of the atmosphere. Once outside the dangers of gravity and air, things get a lot easier for the machine. NASA, of course, has successfully sent dozens of unmanned vehicles into outer space. Galileo has been to Jupiter, the Mars Global Surveyor is still sending data, Pioneer 12 was 6.2 billion miles from Earth when it stopped emitting signals, and Voyager 1 and 2 are now heading out of the solar system. Where the skies are black, no longer is the challenge keeping the spacecraft from falling apart, but keeping the human beings on the inside from falling apart.

Yes, space is a dangerous place. Humans need air, humans need food, humans need water. Typically, these things cannot be found in a vacuum. You do find radiation, however. Astronauts on the Space Station work in low-Earth orbit, beneath the Van Allen belts, where radiation isn't quite as extreme as in naked space. But any interplanetary trip will have to face constant solar wind and periodic solar flares. And if that weren't enough to make a homebody of an astronaut, there are the unpleasant—and unhealthy—effects of microgravity.

Any journey to Mars will likely take six months or more, and a trip to Alpha Centauri, the nearest star, would take at least a generation. For a one-year jaunt in space, a single astronaut needs about 10.6 tons of water— that's about the same weight as the fuel that one of the Space Shuttle's solid rockets uses. To avoid hauling all

STRANGER THAN FICTION

New polymers may allow future spacecraft to morph in flight to adapt to the changing environments through which they travel

A SELF-HEALING SHELL IS JUST ONE of the organs that will make future spacecraft seem practically alive. When a soaring falcon goes in for a dive, it changes its aerodynamics by folding its wings. Spacecraft perform in a myriad of extreme and extremely different environments and need to change as well. Part of the reason the Space Shuttle is so expensive is that its enormous bulk cannot alter itself for the different stages of its flight.

What's efficient for liftoff is not the same as what's efficient for cruising through air at Mach 3 or Mach 8, or orbiting in a vacuum. An efficient spacecraft should look like a rocket at launch, a MiG-25 as it skirts the edge of the atmosphere, and possibly some sort of habitat as it orbits. Such morphing may sound like the technology of a far-away future, but Yoseph Bar-Cohen, a physicist at NASA's Jet Propulsion Laboratory, has already

plastic he's developed that acts like a muscle. When a charge is applied to this polymer, its ions migrate to fit the changing polarity and the whole structure contracts. Bar-Cohen has constructed a four-finger gripper that can grasp objects when it's hit with a current, and a Japanese company has made an extremely realistic-looking plastic fish that swims through charged water. So in a couple of decades, a plane may be able to change its wing shape as the aerodynamic pressure changes at different speeds. Every part of such a craft will have sensors, like a nervous system, that keep its shape constantly on the move. Bar-Cohen thinks that in 50 years, spacecraft may not even be solid. Instead, everything would be made of cells that move to wherever they're needed, and they'll be discarded as they wear out or break. "The whole ship is active, everything is active," says Bar-Cohen. "If you want to sit, your whatever becomes a chair. It's the equivalent of a living creature."

IT'S ALIVE!: Electroactive polymers (EAPs) have become known as "artificial muscles," because of their ability to function like real biological muscles. An electrical charge causes ions to migrate according to changing polarity, and the plastic flexes and contracts. This developing technology holds the promise of spacecraft that can morph according to stages of flight and mission demands. Such technology also promises more playful and science fiction-like "living" materials such as morphing toys, furniture, and especially robots.

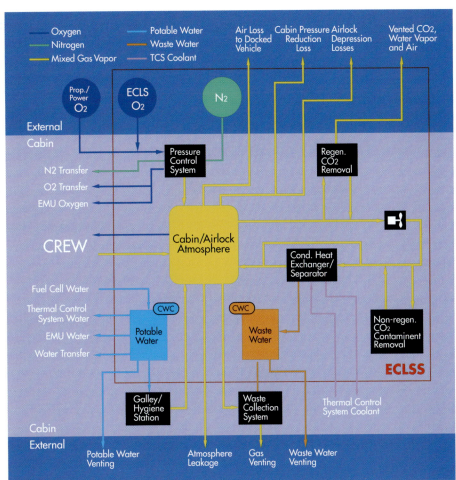

Oxygen
Nitrogen
Mixed Gas Vapor
Potable Water
Waste Water
TCS Coolant

Air Loss to Docked Vehicle
Cabin Pressure Reduction Loss
Airlock Depression Losses
Vented CO2, Water Vapor and Air

External
Cabin

Prop./Power O2
ECLS O2
N2

N2 Transfer
O2 Transfer
EMU Oxygen

Pressure Control System
Regen. CO2 Removal

CREW

Cabin/Airlock Atmosphere

Cond. Heat Exchanger/ Separator

Non-regen. CO2 Containment Removal

Fuel Cell Water
Thermal Control System Water
EMU Water
Water Transfer

CWC
CWC

Potable Water
Waste Water

ECLSS

Cabin
External

Galley/ Hygiene Station
Waste Collection System
Thermal Control System Coolant

Potable Water Venting
Atmosphere Leakage
Gas Venting
Waste Water Venting

AIR AND WATER:
Phase I of the Lunar-Mars Life Support Test Project (LMLSTP), in August 1995, was to demonstrate the ability of a crop of wheat, grown with hydroponics and high intensity light, to provide air revitalization for a human test subject for a 15-day period (left). Phase III of the LMLSTP involved a complicated system for the use and recovery of water, oxygen, and nitrogen (above). Both aspects of the project would be critical to survival on a long mission to the Moon or to Mars.

that H_2O to Mars or beyond, future crews will recycle close to 100 percent of the water they consume. A crew on a trip to Mars would eat mostly pre-prepared, dehydrated, and frozen foods, but the nutritional value of those foods doesn't hold up for more than a few years. For longer trips, astronauts will have to grow their own. And the air will have to be recharged with oxygen from plants. In short, any spacecraft meant to go anywhere farther than the Moon will have to be a miniature biosphere.

To test the possibility of a partial regenerative system for food, air, and water, NASA began the Lunar-Mars Life Support Test Project in 1995. Four astronauts climbed into a 20-foot vacuum chamber known as "the can" and spent 91 days (a new record) in a state of symbiosis with 22,000 dwarf wheat plants.

The wheat plants breathed in carbon dioxide and gave off oxygen, while the humans breathed in the oxygen and let out carbon dioxide. Ten square meters of growing area is sufficient to provide oxygen for one astronaut, and 30 is enough for food (the same plants can work double duty). The wheat has been bred to stay small, but grow fast, and the plants reach maturity in 80 days. That's when the crew harvested them and made bread in their bread machines.

Practically every drop of water in "the can" was used and reused for those

SOME WASTE THAT COULD NOT BE RECYCLED COULD BE MADE INTO BRICKS AND USED TO STOP RADIATION.

91 days. Each crew member had only two gallons of water a day to wash with. They collected all their urine in beakers and dumped it in a recycling system. If a drop got away and hit the floor, it would eventually evaporate and get pulled out of the air by a filter system. Fluids, it seems, are a snap. "We have to make more advances in solid waste," says Donald Henninger, the chief scientist for the project and the head of NASA's Integrity program. "How do you get mass turnover? How do you use it later? It's the highest hill to climb." Some of the waste that simply could not

be recycled could be made into bricks and used to stop radiation. But the real answer to this bulky problem lies in genetically engineering bacteria to do their job more efficiently. "We're trying to duplicate the functions of the Earth," says Henninger. "The Earth has big blobs of orange juice out of the air. So it seems a bit sad that after spending astronomical amounts of money to put people where there's no gravity, we'll have to spend more astronomical amounts of money to artificially induce it.

The many hazards of microgravity

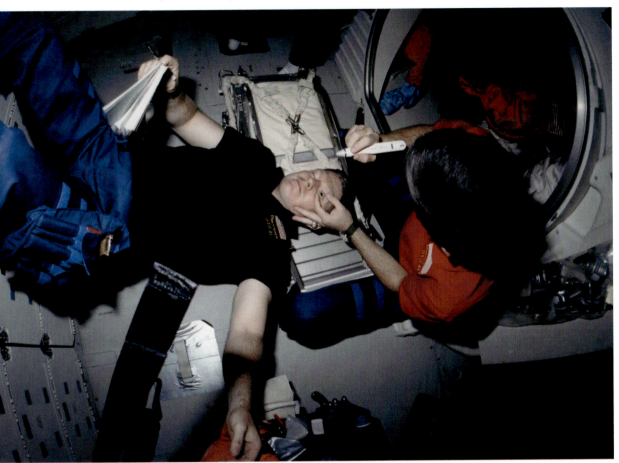

HEALTH WATCH: Astronaut Susan Helms moves a treadmill from the Space Shuttle Atlantis aboard the International Space Station (opposite). Future crews will use it to follow an exercise regimen necessary for health in space. Treadmills can also be useful in other ways: Mission Specialist William Shepherd rests his head on a stowed treadmill while Pilot Robert Cabana prepares to check Shepherd's intraocular pressure. Establishing a database of pressure changes can help evaluate crew health in microgravity.

buffers, like the ocean—we put things in and we take things out. When you have small buffers, it means things are going to change quickly." That means that sensors will have to measure some 500 variables, and an automatic system will need to understand all the subtleties of growth to catch and fix any irregularities before they upset the whole delicate balance.

TO G OR NOT TO G

Half the fun of going to space, one would think, is getting the chance to float about your cabin, stand on the ceiling, and suck

(see page 140) can be mitigated by several hours of exercise a day, but not enough for any astronaut to survive a year-long trek. And astronauts hate exercising in space. The work, for some reason, is more grueling there, and a simple thing like perspiration becomes a disgusting mess—a half-inch-thick layer of sweat undulates on their chests while they work out. Put some spin on the whole spacecraft and the astronaut can get off the stationary bicycle, open a brew, and relax.

It's not that easy, of course. The giant spinning wheel imagined by Werner

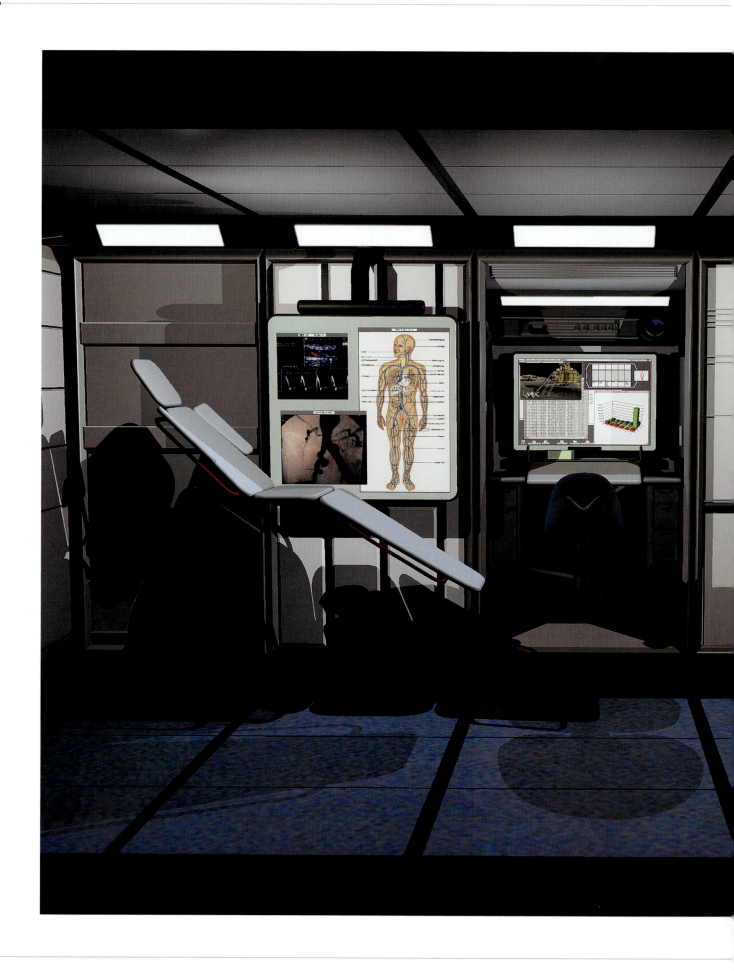

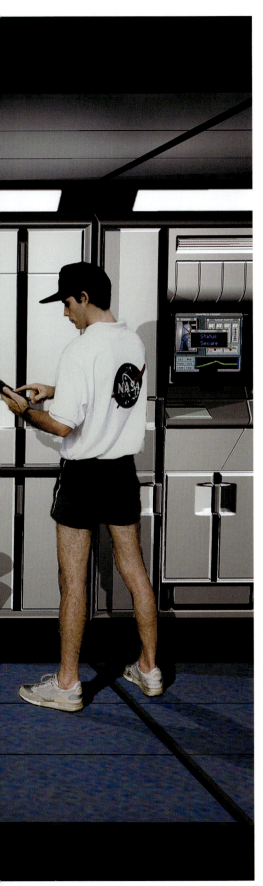

LIFE IN SPACE: A crew member monitors systems on multi-purpose racks in a surface laboratory module (left). In 1952, pioneering rocket scientist Werner von Braun wrote of a 250-foot-wide wheel-like space station (above) made of flexible nylon and designed to rotate to provide artificial gravity.

von Braun, and featured in *2001: A Space Odyssey,* is most likely prohibitively expensive. Even if cost weren't an issue, there are several difficulties that would have to be dealt with. The solar panels of the Space Station take up more than a third of a football field. On the big wheel, those panels would have to have a way of always pointing at the Sun while the rest of the habitat turned. And the "gyroscopic stiffness"—the force that keeps a spinning top upright—would keep the habitat positioned, necessitating even more complex directional changes for the solar panels, and other equipment. Also, the direct, undiffused glare of sunlight in space is disorienting enough for humans, but the long-term consequences of hour-long days and nights of stroboscopic sunlight are unknown.

For projects that are closer to hand than years-long stays in microgravity, like heading to Mars, NASA favors a smaller craft. These may spin on their own or

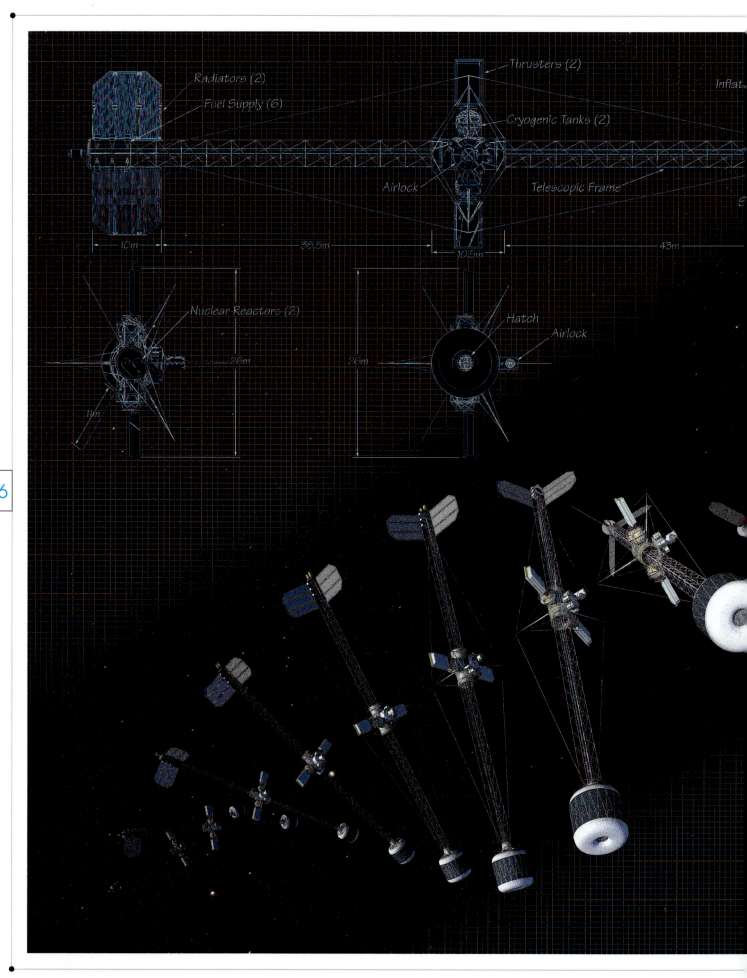

Radiators (2)

Thrusters (2)

Fuel Supply (6)

Inflat...

Cryogenic Tanks (2)

Airlock

Telescopic Frame

10m

38.5m

10.5m

43m

Nuclear Reactors (2)

Hatch

Airlock

26m

26m

1m

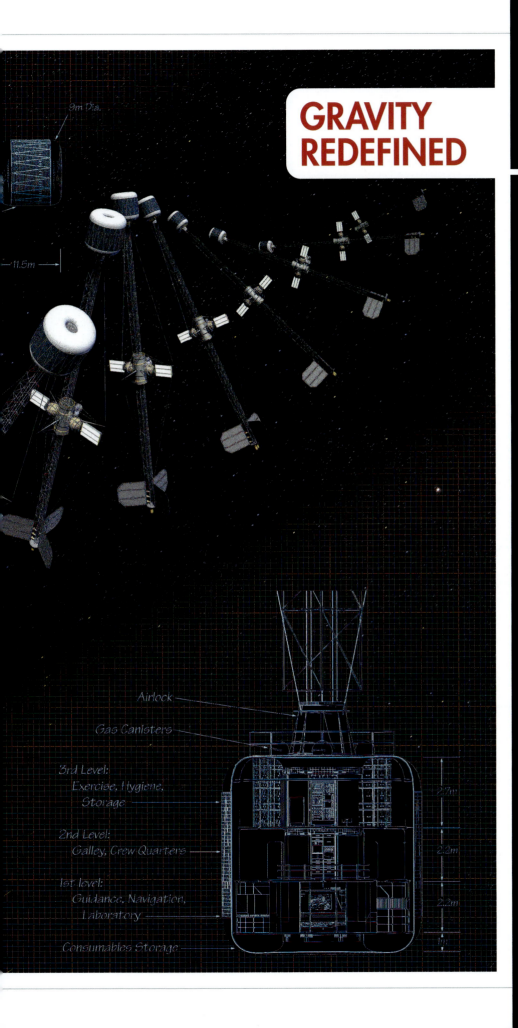

GRAVITY REDEFINED

Crew members travel the vast distances of space in an artificial gravity Habitat, thus reducing the exposure to a zero-G environment. Line drawings show the basic components and dimensions of the Artificial Gravity Nuclear Electric Propulsion Vehicle as well as the interior layout of the crew compartment.

9m Dia.

11.5m

Airlock

Gas Canisters

3rd Level:
Exercise, Hygiene,
Storage

2nd Level:
Galley, Crew Quarters

1st level:
Guidance, Navigation,
Laboratory

Consumables Storage

2.7m

2.2m

2.2m

1m

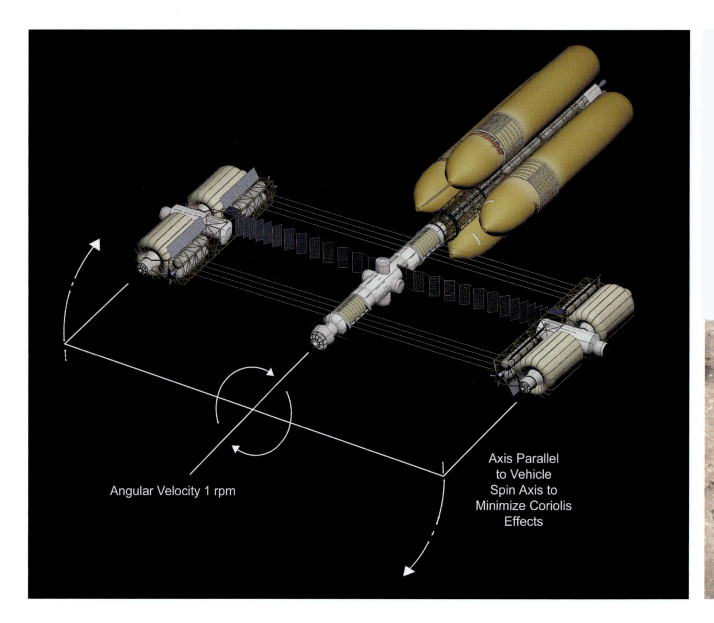

Angular Velocity 1 rpm

Axis Parallel
to Vehicle
Spin Axis to
Minimize Coriolis
Effects

may have an onboard centrifuge where crew members can go for hours-long doses of simulated gravity. Large or small, a spinning vehicle would not provide the same experience of gravity that we get here on Earth. The chief difference will be the Coriolis effect: Walking with the spin will be much easier than walking against it. This may feel like a sudden change in body weight, or it may only feel like the difference between walking up or down a steep hill. Dropped objects will fall in a curve rather than straight down. Turning one's head perpendicular to the spin will cause "cross-coupled

angular accelerations" that make an astronaut feel dizzy and disorientated.

Most of these effects can be lessened by increasing the spin radius. To do that without creating an enormous and exorbitant torus (see page 140) requires a tethered system like the Artificial Gravity Science and Excursion Vehicle (AGSEV) dreamed up by the folks at the Sasakawa International Center for Space Architecture (SICSA). This craft is designed to take 76 tourists and 24 crew members to the Moon and back. On the ground, this spacecraft—a behemoth, no doubt—is not particularly unusual looking. Four

VIRTUAL NEWTON:
A proposed Artificial Gravity Science and Excursion Vehicle (AGSEV) will transport passengers from low-Earth orbit (LEO) to the Moon or Mars and back. The slow rate of rotation and long spin radius (above, left), combined with the orientation of the long axis parallel to the spin axis to minimize Coriolis effects, will make the ride more tolerable. A tether system will allow additional modules to be attached by cables (above, right).

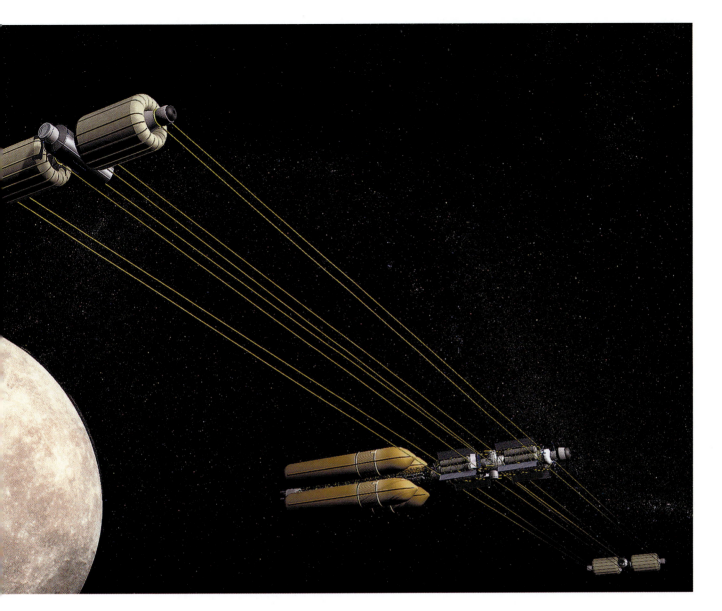

shuttle-style fuel tanks surround the base of a long rocket. At the top of the rocket are two inflatable modules. After escaping Earth's draw, these modules first expand to 27 feet in diameter, then leave the central hub via long "anhydrous glass cables." Solar panels will unfold like an accordion between the hub and the modules. With a minimum of thrust from the modules, the vehicle will spin slowly. To achieve the same gravity as that found on Mars, the spin radius will be 1,140 feet, with the whole craft rotating at one revolution per minute.

With luck—and a whole lot of research—the machines that will bring us to the stars in 100 years will weigh less than we do, will morph to meet our every need, and will cost as much as a pickup truck. Waste will be changed to food by a machine that knows the molecular structure of every known substance in the universe, radiation will be destroyed by super-antioxidants, and gravity will be switched on and off according to our whim. None of these things sound outlandish to the scientists working in those fields today.

We will get there, and cheaply. It's only a matter of time.

Astronaut Mark Lee took a stroll outside the Space Shuttle Orbiter Discovery in 1994.

CHAPTER 6

[HUMANS IN SPACE]

BY WENDY L. SCHULTZ

"FLY ME TO THE MOON, AND LET ME PLAY AMONG THE STARS; LET ME SEE WHAT SPRING IS LIKE ON JUPITER AND MARS"

—BART HOWARD
(SUNG BY FRANK SINATRA)

DIARY, 4 JULY 2103: PARTICIPANTS IN THE THIRD ANNUAL OLYMPUS MONS SUMMIT RUN ARE CHATTING AND CARBO-LOADING AT RED HORIZONS, THE MICHELIN-STARRED RESTAURANT ON THE RIM. PHOBOS AND DEIMOS ARC OVERHEAD, BRIEFLY IN SYNC, AS THE WINDOWS

STEPPING STONE: A Station at Earth-Moon Lagrange point L1 might serve as a platform for trips to the Moon (above) or to other planets.

flare rose-gold, and the sloping landscape reddens in the setting sun. You connect to the Interplanetary Internet and send a snapshot to your grandparents on the Moon: You know they'll relate to your grin, because all they could talk about when they got back from Mars was the lingering sense of wonder and delight. Well, that and how good one-third gravity felt to their arthritis—that's why they moved to Habitat Tranquilitatis last year. Yeah. *And* the low-G wing-gliding. *End entry.*

WHY MOVE TO SPACE?

Popular responses for why humans should move to space vary: Life naturally expands; humans naturally explore; overcoming challenges spurs creativity and technological innovation; space is abundant in energy and raw materials; having a human foothold elsewhere in the solar system creates a genetic reservoir as a back-up in case of asteroid "extinction events." As many reasons exist to go to space as there are people who dream of the journey.

Biology drives us, beauty calls us, and potential profits beckon us. Proponents of the space program cite the economic engine it creates. Albert A. Harrison, author of *Spacefaring: The Human Dimension*, reminds us that "every dollar spent on the Apollo Moon program translated into seven to eight dollars returned to the economy in new goods and services." In addition to the economic multipliers derived from a large space sector, enthusiasts count on technical innovation creating new products and markets. But the economics of space are inevitably the economics of gravity wells. The adventure may generate innovations, but first we will need innovations to create the infrastructure needed for the adventure.

WHAT'S OUR PLAN FOR EMIGRATION?

So what are our options for scrambling out of our gravity well? In NASA's early days, Werner von Braun proposed a stairstep approach to coloniz-

COULD WE ESTABLISH BASES ON THE MOON? COULD WE MAKE MARS A GARDEN PLANET IN THE NEXT CENTURY?

ing space: First, master manned space flight; next, build a space station for use as a construction platform and staging ground for putting a base on the Moon. Once a Moon base proves successful logistically, technologically, and socially, launch an expedition to Mars. In the 1970s, Gerard K. O'Neill, a Princeton physicist, suggested that popping our heads above one gravity well only to hunker down at the bottom of another one was lunacy. Rather than colonize planets, O'Neill proposed perching rotating orbital settlements at Lagrange points, balanced between Earth and Moon, and scooping up the bright incoming energy radiating from

the Sun. Don't "colonize" any planetary territory at all: Build your own! Best of all, building orbital colonies is excellent practice for building interstellar generation craft: Think "orbital colony as RV."

In the 1980s and '90s, shuttle costs rose and the proposed Space Station Freedom design was scrapped, redesigned, rebudgeted, re-staffed with an international team and cross-cultural technical standards, and renamed the "International Space Station" (ISS). Meanwhile, a new paradigm shook itself loose from the von Braun stairstep. Vigorously championed by Robert Zubrin, "Mars Direct" suggested sending explorers straight to Mars, sidestepping little details like space stations and Moon bases. Automated supply craft would precede the crew, ensuring a backup return vehicle, and would refine fuel from chemicals on the Martian surface. This idea fired the imagination of many who felt cheated by the collapse of the Apollo program.

Administrative momentum and invested effort carry the ISS forward—accompanied by numerous small, innovative planetary probe missions. Against a stunning backdrop of Martian landscapes beamed back by Pathfinder, the Mars Surveyor, and Odyssey, the space colonization debate rages on. How quickly could we establish scientific bases on the Moon? Could we make Mars a garden planet in the next century? Should we? Or should we design our own biospheres and plant them in O'Neill colonies?

WHAT CHALLENGES WILL WE FACE?
BASIC NEEDS

Konstantin Tsiolkovsky (1857–1935), the Russian pioneer of astronautics, likened Earth to a cradle. Cradles are cushioned and warm, and come with

intelligent, adaptive food suppliers who also dispose of our waste. Whether that adaptive system is Mom or Gaia, she does not follow us outside as we make our way in the universe. Thus the immediate challenge to living in space is meeting basic needs: coping with vacuum, 500-degree variations in temperature, inimical levels of radiation, and

keeps launch costs low, it is less than elegant. In contrast, closed biosystems—greenhouses in space—offer a more organic approach, but are complex, heavy, and large: NASA's prototype Closed Environment Life Support System (CELSS) requires 12 square meters of plantboxes to feed one spacefaring vegetarian. Russian space scien-

lack of food or potable water. When we scramble out of this gravity well, we must carry our biosphere on our backs. People need, at minimum, four and a half kilograms of oxygen per day, 15 to 20 kilograms of water per day, and just over three kilograms of food (dry weight) per day.

The brute force approach to supplying these needs is, indeed, to pack them up with us in air tanks, water tanks, and vacuum-packed meat-loaf . . . with extra hot sauce. While this works, and

tists pioneered these manned, closed ecosystems in 1965; NASA began testing the concept in 1989's "BioHome" and ran 90-day Lunar-Mars Life Support Test projects between 1995 and 1997. A properly designed "greenhouse in space" would use plants to generate oxygen, absorb carbon dioxide, purify water, and recycle solid wastes. While inefficient in terms of mass and space, a greenhouse would provide an oasis of life in an otherwise sterile technical environment.

MANNING MARS: Zubrin's Mars Direct Mission (MDM) called for two Habitat Landers—one primary Lander and a backup—and an unmanned Earth Return Vehicle (ERV, far right). The ERV would land first, bearing liquid hydrogen, compressors, a chemical processing unit, several rovers, and a small nuclear reactor to lay the groundwork for human habitation and for the return home.

PREPARATIONS FOR THE ASCENT

Would-be space travelers train in a variety of extreme environments in preparation for a very different kind of life

NASA ANALOG SITES: ANTARCTICA AND AQUARIUS

Want to practice your interplanetary field exploration skills? Head to Antarctica. Like Mars, Antarctica is a polar desert, averaging less than 5 cm of precipitation every year, with temperatures varying from -50°C to 10°C. Mars is even drier and colder, with less than 1 cm of precipitation annually and ground temperatures ranging from -107°C to -17°C. Antarctica's Dry Valleys even look like Mars. Ice-free, their stark landscapes of wind-scoured rock and dry riverbeds offer no obvious promise of life. Consequently, optimism regarding the possibility of finding microorganisms alive on Mars rose when microbiologist E. I. Friedmann found both lichen and cyanobacteria thriving, tucked amid crystals of sandstone in these arid oases.

Supplying Antarctic stations with food, fuel, and other necessities of life—toilet paper! aspirin! toothpaste! socks! chocolate!—requires a secure logistical chain, as will any space habitat in its early days. Space habitat designers draw lessons from crew compositions, shopping lists, and resident activity patterns at McMurdo Base and other Antarctic stations. Boredom, they found, causes stress more often than personality clashes.

But what if you want to practice working in orbital habitats rather than exploring planets? Since October 2001, teams of four "aquanauts" at a time have trained for space missions below the waves off Key Largo. Their analog space habitat is Aquarius, the world's only undersea laboratory, operated by the University of North Carolina–Wilmington. Transit to Aquarius requires only a leisurely swim in scuba gear, but returning home is a bit trickier: Aquanauts must undergo 17 hours of decompression for any stay exceeding 80 minutes. This compact habitat maintains an ambient pressure of 2.5 atmospheres. As a result, Aquarius can keep a wet porch open to the ocean, through which aquanauts can exit for unlimited work time at their bottom depth of 19 meters.

WIDE OPEN SPACE: The Mars Society has established what it calls "analog" field exploration stations for testing equipment and training space travelers. The sites are in extreme-climate locations like Antarctica, the Arctic, and the Utah desert, and more are planned in Iceland and Australia. Training on Utah's planet-like terrain (left) began in January 2002.

With only 11 cubic meters of living and laboratory space, it's not the lap of luxury—but it is about the same space as the ISS habitation module. So the teams get a very good feel for performing complex scientific and technical tasks while separated from mission control, and living elbow-to-elbow. The underwater environment also allows them to practice construction techniques in weightless conditions, simulating space station assembly procedures.

MARS SOCIETY ANALOG SITES

Getting us up and out of the gravity well and on to other planets does not rest on the budgets of the government alone: Private initiatives exist to push humanity aggressively toward Mars. One such is the Mars Society, which has established its own Mars analog field exploration stations. In 2000, they constructed the Flashline Mars Arctic Research Station on Devon Island in Canada, following on the heels of NASA's Devon Island Arctic research station near the Haughton impact crater. Flashline first simulated Mars operations for two months in 2001, and has returned every year since. In 2001, the Society also started developing an analog pressurized rover, and an additional analog site in Utah. The latter began operations in January 2002, which also saw the Society planning for additional sites in Iceland and Australia for 2003. Devon Island serves as both a climatic analog and, with the Haughton impact crater, a landscape analog; Utah provides an opportunity to test equipment before deploying it in harsher environments; Iceland's geothermal activity mirrors possible sites where life might be found on Mars; and Australia offers embedded fossils whose ancient traces allow the exercise of detective skills Mars explorers will need.

For the next 20 years or so, these sites and others—for example, high altitude sites offering analogs for Mars's thin atmosphere—will serve to train space explorers. By the middle of the century, they may well be proving grounds for potential colonists, allowing candidates to test their mettle when faced with isolation, extreme climatic conditions, and demanding physical and intellectual tasks.

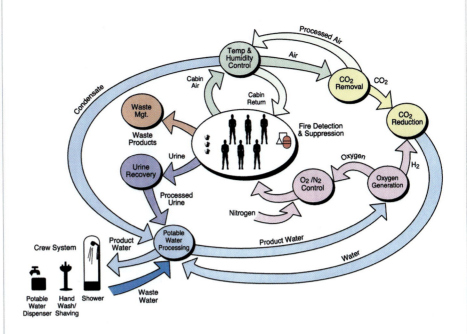

GO WITH THE FLOW:
The Environmental Control and Life Support System (ELCSS) will recycle needed resources on the ISS and offer a model of what will be required for space exploration.

And space need not be an issue in space. As Constance Adams's prototype TransHab design demonstrated, a living module suitable for the ISS or an initial planetary base can provide both private and communal space as well as shield residents from radiation by clever placement of water reservoir tanks. Adams's prototype uses flexible materials combined with a locking endoskeleton, allowing it to be flown into orbit folded; when pressurized, it blooms into a 340-cubic-meter space home. The Kevlar-Nextel-Combitherm-Nomex composite skin, padded with foam, is stronger than metal and protects residents from particle impacts. Combined with sophisticated biosphere design, such creativity in space architecture represents the first step toward construction of the orbital colonies first envisioned by O'Neill, and championed today by aficionados of Lagrangian space settlements.

SIDE EFFECTS

After meeting our basic needs for air, shelter, and food, we must counter side effects generated by the hypergravity of take-off and the microgravity of

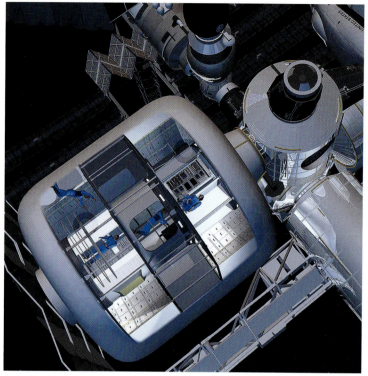

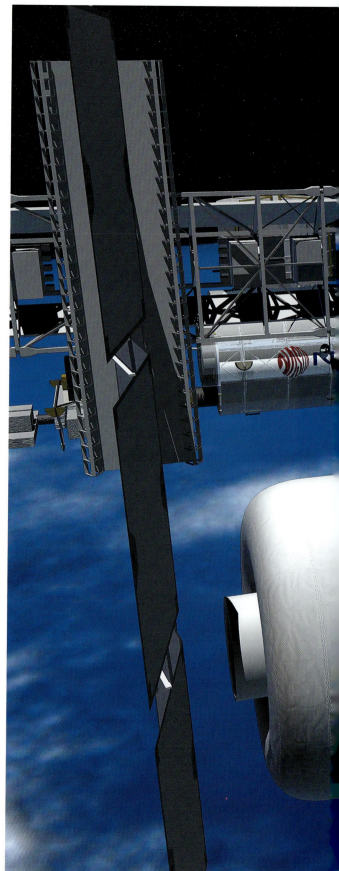

HOME, SWEET HOME:
Transhab Modules (cutaway, above), as might be attached to the ISS (opposite), would provide a safe, homelike environment with artificial gravity, body-friendly exercise devices, and psychologically soothing videos and sounds of Earth. Such a cocoon, though small, comprises three levels—a lower wardroom level, a middle level with living quarters and mechanicals, and an upper level with an exercise room and shower. Inhabitants would be protected by a multi-layered shell of scuff barrier, bladders, kevlar, shielding, and thermal blanketing. A tunnel at the top provides entry to the rest of the space station, for work and interaction.

extended flight in free fall. On launch, acceleration can equal 3.2 Gs, and deceleration on re-entry can reach 1.4 Gs. Granted, extreme rollercoasters subject riders to as much as 4.5 Gs, but only for six seconds; shuttle acceleration on launch lasts for eight minutes. Aerobraking into Martian orbit will produce deceleration equivalent to five times Earth's gravity. In hypergravity, bodily fluids weigh more, stressing the heart as it pumps blood to the brain. Inefficient heart response could cause space travellers to "gray out," or in the worst case, to pass out. NASA's Malcolm Cohen suggests that pre-conditioning would help. Prior to lift-off, space passengers could add centrifuge sessions to their daily work-outs in order to help their hearts adapt to fluctuations in acceleration. Aerobic conditioning, stretches, push-ups, sit-ups—and spin-outs.

Only a handful of people—all men—have lived in space for over 10 months. We cannot base our designs for extended human habitation of space on

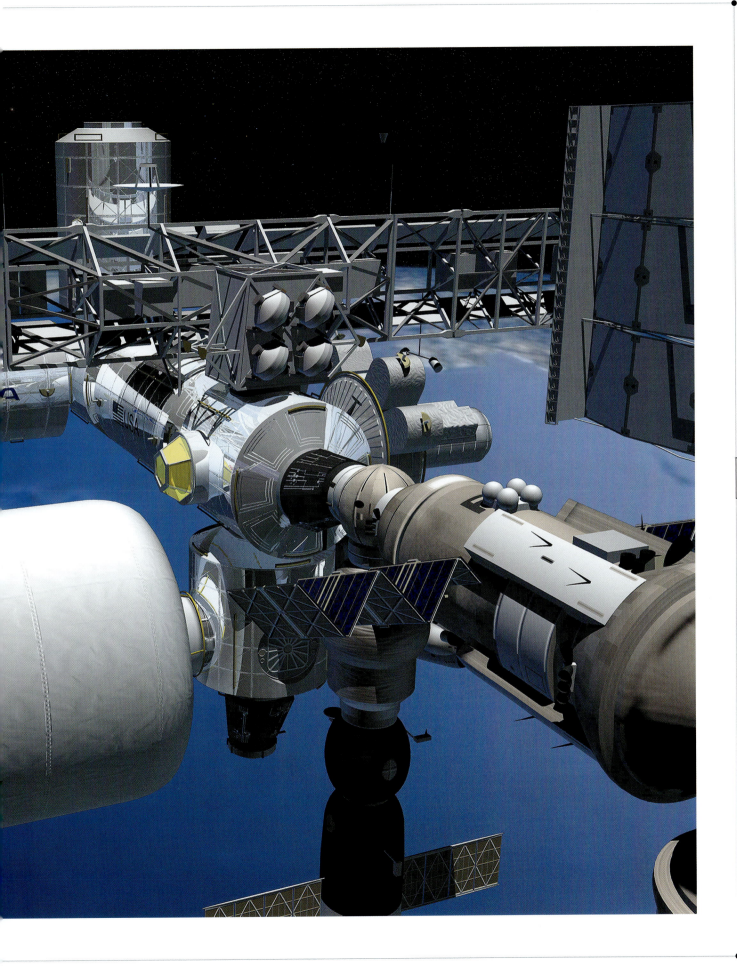

this small a sample. But the experience thus far *has* demonstrated microgravity's effects on human biology. First, swollen feet are never a problem on long spaceflights: Fluid pressure equalizes, and over time, the body's fluids move up to the head. Space travelers often develop "space sniffles," and look much rounder-faced. Meanwhile, because the brain thinks the entire body is being flooded with excess fluid, it triggers fluid excretion, which decreases blood volume, compounds calcium loss, and dehydrates the body: Space travelers

spending over two hours a day on a bike or treadmill; "negative pressure" suits, which draw fluids back down to the lower half of the body; boots, which press on the soles of the feet, simulating the feel of standing; flight deck and station design features, which distinguish between "floor" and "ceiling." Given that increasing muscle mass increases bone mass, short bursts of strength training, combined with extended aerobic workouts, may help maintain musculo-skeletal health. These, however, are clumsy palliatives;

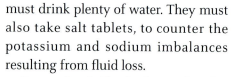

BONES DECALCIFY, MUSCLES ATROPHY. . . . STILL THINK A CAMPING TRIP TO THE ASTEROIDS SOUNDS LIKE FUN?

must drink plenty of water. They must also take salt tablets, to counter the potassium and sodium imbalances resulting from fluid loss.

Bones decalcify, the heart shrinks, muscles atrophy, immune system T-cells show damage, smell and taste are degraded, and flatulence increases. Most space travelers also experience vertigo and disorientation. "Space adaptation syndrome"—motion sickness—results from a mismatch between the inner ear's balance signals and visual signals. Still thinking an extended camping trip to the asteroids sounds like fun?

In-flight countermeasures abound:

advances in both space engineering and bioengineering will offer more fundamental solutions.

While physicists search for a Theory of Everything, generating real gravity artificially will have to wait. But simply rotating our space vehicles and habitats could create a facsimile of gravity via centrifugal force. In 2005, the student-designed Mars Gravity Biosatellite will fly two generations of mice in a centrifuge to LEO to test physiological response to rotating habitats. Engineering an interplanetary craft to rotate while accelerating toward its destination, and decelerat-

HORN OF PLENTY: Whole settlements in space require a much more ambitious approach than ISS-attached individual Transhab modules. A 1975 NASA design study, done in conjunction with Stanford University, proposed the Stanford Torus, a donut-shaped ring one mile in diameter, which could house 10,000 people and would rotate once per minute to provide Earth-normal gravity.

ON THE SCENT

The smells of home may be just the tonic for sensory-deprived space dwellers

"WHAT DOES THE SPACE STATION smell like?" During a June 2003 interview with schoolchildren, International Space Station commander Yuri Malenchenko and science officer Edward Lu found themselves fielding this question. Tactfully, they replied that the ISS smelled sterile, of machinery. But Commander Malenchenko then described the docking of the Russian supply ship: When they opened the valves equalizing pressure, the scent of green apples flooded into the ISS. He smiled as he described it—clearly it was a pleasant memory for him.

Scent strongly evokes memory—this powerful and well-documented phenomenon has been labeled the "Proust effect," based on the author's vivid description of it. The Proust effect could help ensure our psychological well-being in space habitats, by grounding us emotionally, and by connecting us to the familiar. While the human brain devotes much less processing power to scent analysis than any other large predator, we are nonetheless accustomed to immersion in a living biosphere, rich in fragrance and smell—some pleasant, some unpleasant. Also, EEG studies indicate that scents do affect brain function: Ylang-ylang stimulates alpha waves, a sign of relaxation, and rosemary depresses them, a sign of stimu-lation. Unfortunately, one person's fragrance nirvana is another's nausea, so using scents in space to enhance learning, working, or relaxation would require careful control.

Astronauts may adapt to living in a sensory-poor environment for their missions, but they know they are returning to Earth. Permanent human habitation in space will require cushioning our living environments with stimulating textures, pleasing colors, familiar scents, vivid tastes, and both pleasant soundscapes and the possibility of blissful silence. The latter is currently lacking in the ISS; engineers have struggled with noise reduction throughout its design. In the next century, interior design companies for space habitats will team graphic artists, ecologists, bioengineers, psychologists, nanotech programmers, anthropologists, and, perhaps, "noses"—just as NASA employs "chief sniffer" George Aldrich and his team of chemical specialists at the Molecular Desorption and Analysis Laboratory to smell test any new supply sent to the ISS before it reaches orbit. Because in space you can't open the windows.

THE NOSE KNOWS: Smells—pleasant as well as noxious—have a greater psychological impact on our daily lives than many of us realize. Addressing the olfactory deprivation of long-term space travel will be crucial in the planning of future space environments.

ON THE TREADMILL: A daily regimen of two-hour exercise sessions is critical to counter-balancing the pernicious effects of extended space travel. In microgravity, one loses bone density, muscle mass, body fluids, and blood volume, not to mention suffering fom T-cell damage, vertigo, and nausea. The spartan surroundings on Atlantis, Orbiter Vehicle 104 (right), will in the future give way to a more user-friendly and psychologically reassuring environ-ment (below), which will include the sights and sounds of Earth in order to keep travelers oriented, content, and "grounded"—as well as physically strong and healthy.

ing into orbit around it, is no mean feat. Rotating a stationary habitat is a much simpler proposition. Choosing to live on planets is simpler still. But colonizing the Moon or Mars still means constructing radiation shield-ing—or opting to live underground. Studies show lava tubes below the sur-face of the Moon or Mars would ade-quately protect people against cosmic rays and solar flares, not to mention UV radiation. But who wants to ride a rocket to Mars and then live in a hole?

Over the next century, bioengineer-ing and nanoengineering will converge. In doing so, they will radically trans-form how we interact with matter. We could supersede the need for radiation shielding by simply restructuring our bodies. By 2100, spacefarers might rou-tinely undergo biochemical and genetic modification to "harden" them against radiation damage.

In the next decade, space industry

BEND IT LIKE BECKHAM?

Future space travelers will find their games fundamentally changed

IMAGINE: YOU'VE JUST RETURNED from a one-week visit in space. You suffered two days of Space Adaptation Sickness—disorientation, nausea, loss of concentration, sleeplessness, and vomiting—and lived to enjoy the freedom, grace, slapstick humor, and delight of micro-gravity. Sipping reflectively on your soda, you set it aside to jot down a few thoughts on the experience in your diary—only to realize that on Earth, soda cans don't float conveniently in mid-air, and you've got a spill to mop up. Astronauts say such lags in situational adaptation are common. What other errors in expectation can we anticipate for our future lives in space?

The Intra-Orbital Baseball Association will have a difficult time in the 2100 New Worlds Series: In orbital habitats, rotating to provide a simulacrum of gravity, all balls will be curve balls. Or at least, they will appear to be, even while actually traveling in a straight line. Why? Because the rotation of the habitat will fool our eyes into seeing a curve as the background moves relative to the trajectory of the ball. Want to practice acclimating as an intra-orbital umpire? NASA scientist Al Globus and Stanford University researcher David Whitney offer a clever applet, Ringworld 1.0, that lets you experiment with dropping objects in a rotating cylinder (see the applet in action at http://lifesci3.arc.nasa.gov/ Space-Settlement/teacher/materials/ ringworld/).

Any FIFA Worlds Cup soccer team from Earth would have similar problems adjusting to play in the atmospheric conditions of Mars, or any habitat with a similarly thin atmosphere. Why? Because their curveballs would all curve away from the direction of their spin—here on Earth, soccer players, table tennis players, and golfers intuitively expect balls to swerve into their spin. In thick air, the pattern of molecules colliding as the ball spins pushes it into the spin—in thin air, the distance the molecules travel before colliding is larger than the diameter of the ball, and they push the ball away from the spin. Of course, no one's going to play soccer in the thin, poisonous, particulate-filled air of Mars. But it's not unlikely that space colonies—whether planetary or orbital—that manufacture their own air would economize with air pressures at levels higher than Mars, but lower than Earth. At intermediate air pressures, suggest scientists at Stockholm's Royal Institute of Technology, airflow effects around a spinning ball cancel out: A curveball . . . won't.

No matter what your favorite sport—or game, or dance, or art—may be, the widely varying environmental contexts that outer space presents will make almost all leisure pursuits extreme sports by 2100: swimming in microgravity, à la John Varley's short story, "Blue Champagne"; marathon running around the world, à la Kim Stanley Robinson's *Green Mars*. Living in space will mean playing in space, and endless innovation in sports and the arts.

SPACE BALLS: Differences in air and gravity will dramatically alter the way in which our baseballs and soccer balls behave.

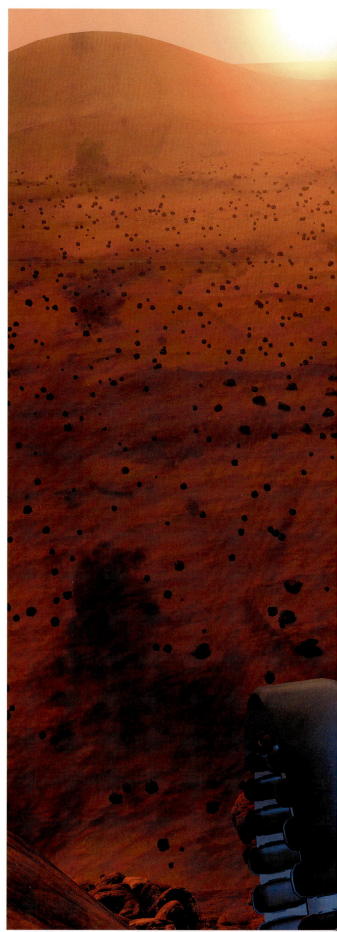

work teams of humans directing "robo-nauts" will be common. But we may also be able to "harden" ourselves against both hyper- and micro-gravity. Designers are already creating "wear-able computing," and researchers are already growing nerve cells on silicon chips. "Bioartificial" organs—mixing mechanical and biological compo-nents—are currently in clinical trials. By 2100, the boundary between a bioar-tificial, networked, semi-autonomous spacesuit and its wearer will blur beyond recognition. The resulting flex-ible exoskeletal skin will cushion its owner from environmental variations of atmosphere, acceleration, radiation, and impact.

SOCIAL PSYCHOLOGY

"All the conditions necessary for mur-der are met if you shut two men in a cabin measuring five meters by six and leave them together for two months."

—*RUSSIAN COSMONAUT VALERY RYUMIN.*

Building team cohesion by uniting against irritating micromanagers is not the best practice. But it has happened in space: A Skylab crew once shut up ground controllers by simply shutting

IRON MEN: Astronaut Nancy Currie (above) shook hands with a Robonaut prior to a testing session in 2003. A mobile, autonomous modular robot (right) would enjoy a greater range of exploration on hazardous terrain than its human counterpart.

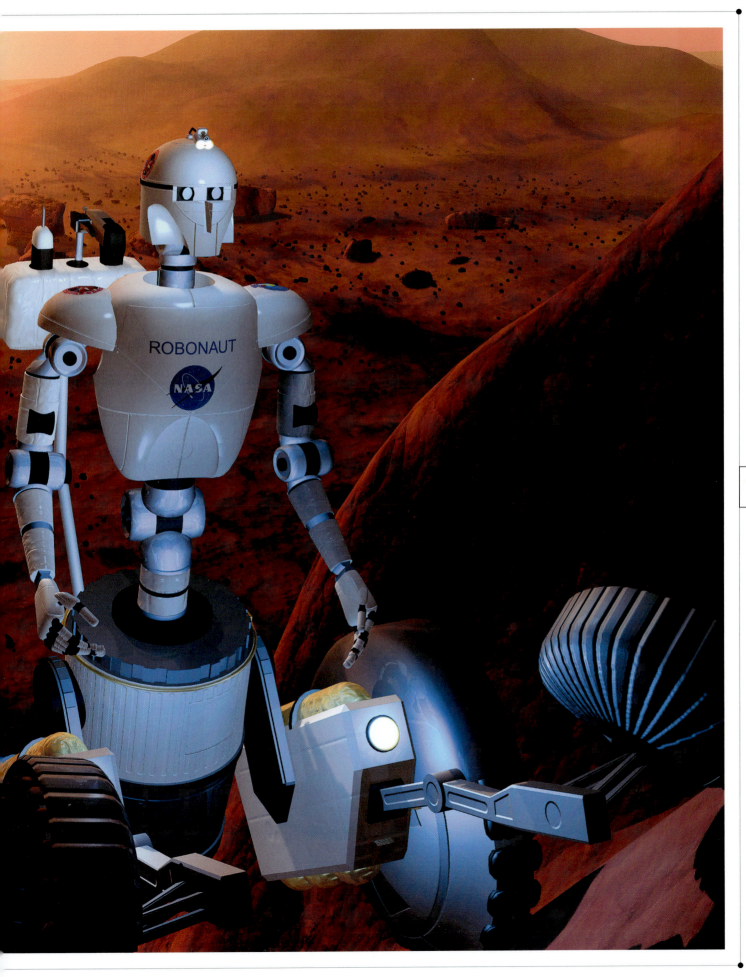

off their radio. Yet NASA has historically been culturally averse to questions of psychology—perhaps as a holdover from test flight days, when pilots were wary of being grounded by any medical authority.

Much is changing. Space explorers

ground controllers create therapeutic surprises to keep crews lively and happy: transmitting the sounds of Earth—waterfalls, rain, urban street sounds—as well as audiovisual hookups with family and friends. NASA creates digital "photo albums"—

I'M FROM VENUS, HE'S FROM MARS: How will the men and women of the future get along in space? The rigors of extended travel will make it difficult to maintain the apparent camaraderie exhibited in 1997 by the crew of the Space Shuttle Discovery (left to right): Kent Rominger, Robert Curbeam, Stephen Robinson, Curtis Brown Jr., N. Jan Davis, and Bjarni Tryggvason.

will never be drawn from the ranks of the risk-averse, but they can no longer be drawn from the ranks of the risk-addicted. The psychological challenges of long team missions include both isolation and lack of privacy, monotony and boredom, overwork, and cross-cultural and gender tensions, to name just a few. Both Russian and American

including family pictures, musical selections, and video clips—with a different entry for every day in orbit.

Team analog training at isolated sites like Antarctica or the Aquarius also helps teach mission crews what to expect psychologically of themselves and their teammates in terms of problem-solving, personality style, and

response to stress. In addition, NASA scientists are researching and designing expert systems to coach mission teams through crises: primitive versions of the *Star Trek* "holo-doctor" expert system. These systems will identify stress and depression by monitoring acoustic changes in speech patterns and muscular changes in facial expression. An interactive module will then coach crewmembers in self-assessment and self-treatment of depression.

The strategic communication technologies for space colonization will be the "soft" technologies of group process: facilitation, mediation, and alternative dispute resolution. Just as the last 50 years brought innovations in telecommunications and computing, it also brought innovations in meeting process: out with Robert's Rules of Order; bring on nominal group technique, facilitators, agendas, groundrules, and groupnotes. During long space voyages, effective collaboration despite cultural and gender differences will be critical to survival.

For extended space expeditions, NASA will also need to get over its American prudery with regard to sex, and acknowledge its potential impact on crew dynamics. For this very reason, one Russian space scientist has suggested sending all-male crews to Mars, to avoid brawls. On the other hand, an American space scientist has asked whether an all-woman crew might not be the best choice for a Mars mission—not only would they mass less, they would be more likely to talk through their disagreements than to fight about them.

SOLAR SOCIETY IN 2100

The more astronauts, scientists, and space immigrants use group facilitation tools to accomplish joint tasks, problem-solve, plan, and create, the more those tools could evolve toward a space-based governance system. When combined with the Interplanetary Internet—the satellite-based digital network designed to expand the Internet to Mars and beyond—such a culture of participatory decision-making could be the seed of a solar system-wide electronic democracy. As the original Internet did, the Interplanetary Internet could morph from its initial purpose of scientific and educational communication, through storytelling and entertainment, to citizen debate and voting.

Diary, 5 July 2103: historical notes. We

> NASA WILL NEED TO GET OVER ITS AMERICAN PRUDERY AND ACKNOWLEDGE THE IMPACT OF SEX ON CREW DYNAMICS.

have populated our solar system with silicon-based agents—some stupid, but most intelligent and self-programming. The more we spread micro-robots, smart sensors, and nanotechnological devices throughout the solar system, the more we nurtured an interplanetary ecology of active, ambient intelligence. At that point our challenges became moral and ethical, rather than technical. Who builds the habitats? How are they governed? Who decides whether to terraform Mars? Who owns the asteroids? Can advertisers erect billboards on the Moon? Will Starbucks open

THE FIRST MARTIANS

The first genuine colonies on Mars might include biodomes to be used for initial terraforming experiments prior to their execution planet-wide. Such a colony might accommodate several hundred to a thousand people.

a coffeeshop on the summit of Olympus Mons? Will all of Mars be opened to prospectors and colonists, or declared a geological treasure for all of humankind—and spacekind?

We spread ourselves out as well: Once we mastered nanoassembly of space settlements and their internal biomes, people from every culture poured off the planet in the great diaspora of 2055, and the age of Space Separatism began. O'Neill orbitals were built for every religion, philosophy, sect, and culture you can name. But extreme homogeneity is not healthy for plants or people, and most colonists—both planetary and orbital—layered their identities across multiple interplanetary interest groups, if only virtually. Where the fad once was cultural retrieval—signing up to live in a colony where you could immerse yourself in the language and worldview of your ancestors—it shifted, as we approached the turn of the century, to focus more on diversity. When we weren't arguing mineral rights, of course.

We haven't used fossil fuels in 50 years, and we harvest plastics from bioengineered plants—our "refineries" now resemble farms and gardens. So are we installing new factories and production facilities on the Moon and Mars, or terraforming them? The ethical debate rages on: The Reds—Martian environmentalists—wish to preserve Mars as is. Paraterraformers, focussed on roofing over half the planet with a "worldhouse," argue that they are practicing "conservative, contained terraforming," by keeping it all under a dome. The Reds are threatening deconstructive nanotech unless the project stops.

News flash: The Stellar Gypsies are now inflating the magnetic "Winglee balloon" around their O'Neill habitat, with the celebration concert starting in five minutes. They wish to sail to a nearby star on the solar wind. Approximately 65 percent of them will enter wintersleep—the artificial hibernation we learned to induce by studying bears—to avoid monotony-induced stress, but many families and clans will make the voyage a generational project. *End entry.*

RACING TO SPACE

The Columbia tragedy has further delayed ISS construction. Meanwhile, exploration of the Moon and Mars by robotic probes and rovers continues. The European Space Agency (ESA) points out to its public that the cost to Europe of participating in the ISS, when amortized over 30 years and distributed among the sponsoring nations, is equivalent to one cup of good coffee per European per year. For Americans, that probably works out to a few pots of coffee per year. And if the Old Occident chooses not to subsidize our move into space, the New Orient just might. Waiting in the wings are China and, potentially, India.

The Chinese are pushing ahead with their own space program, creating the Shenzhou ("divine vessel") for manned space flight by revamping Russian Soyuz and spacesuit designs. Given their human resources and potential for economic development, the first permanent colony in space is as likely to speak Mandarin as English, and the Middle Kingdom would at last truly find itself suspended between Earth and the far reaches of heaven.

By the middle of the 21st century, India will overtake China as the world's most populous nation. Businesses around the world already rely on India's college-educated labor force, outsourcing their engineering, programming, and research needs to Indian consulting companies and call

centers. It takes no very great leap of social imagination to envision that intellectual power harnessed to solve the technological challenges of reaching and exploiting the raw materials of space. Perhaps Hindi will be the first language of spacekind.

Or we could see the decentralization

force the speed of innovations pointed toward space. The more low-cost strategies emerge, the less likely it is that these enthusiasts will wait for cumbersome government agencies to get us out of the gravity well. But the resulting feed-forward effect also implies that the farther we get into space, the more we

EARTH ON MARS: An immense greenhouse, covering several thousand hectares at the Martian equator, could provide suitable Earth-like temperature and atmosphere to enable plant growth. Such a structure would be necessary, given the -55°C mean temperature and the air composition of about 95% carbon dioxide. It could also provide housing for about 250,000 intrepid settlers.

and privatization of human endeavours in space. From millionaires to popstars, visitors are making their way into orbit.

After all, the next century will see genomics and nanosciences merge. For good or ill, that synchronicity will produce explosive change. The demands of scientists, entrepreneurs, explorers, extreme sport enthusiasts, and resource-hungry societies will only rein-

will transform our tools and ourselves with nano- and bio-engineering, and the more each successive generation of spacekind will diverge from their counterparts on Earth.

Whatever the language of the first colonists, all others will follow, as humanity seeks its potential in the challenge of the stars. Welcome to generation zero of spacekind: *Namaste.*

RUNNING INTERFERENCE: The 2009 Space Interferometry Mission will use wavelengths of light to gain more precise measurements of the distances and positions of stars and planets.

CHAPTER 7

[SEARCH FOR LIFE]

SMALL BLUE PLANET (SBP) LOOKING FOR CONVERSATIONAL COMPANIONSHIP. WILLING TO TRADE PERIODIC TABLE OF THE ELEMENTS, MATHEMATICAL ASSUMPTIONS, AND MOZART. BIG RADIO TELESCOPE OR FAST, POWERFUL PROBE A MUST. WORLD-CONQUERING FLESH-EATERS NEED NOT APPLY.

LOOKING FOR LOVE IN ALL THE WRONG PLACES?

"WHERE IS EVERYBODY?" PHYSICIST ENRICO FERMI FAMOUSLY ASKED. IMAGINING LIFE ELSEWHERE IN THE

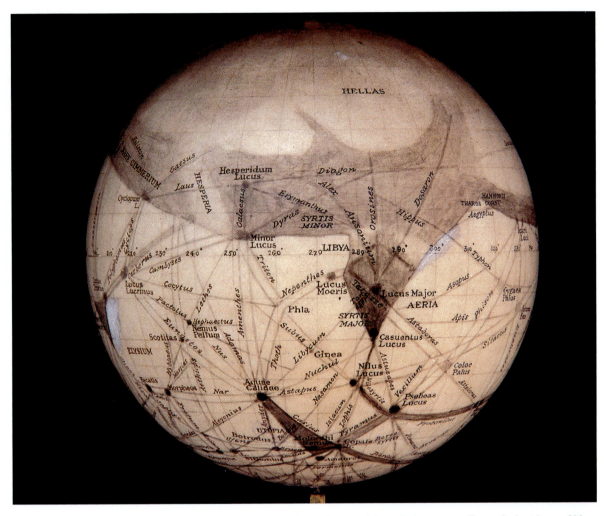

MAP QUEST: Around the turn of the last century, amateur astronomer Percival Lowell mapped the Red Planet in an effort to find evidence of life.

universe has been a human pastime since we invented language. We are primates, after all, and yearn for social connection. Galileo shattered the hierarchies of heaven when he spotted moons orbiting Jupiter, and tore us from our comfortable niche just below the angels in the Great Chain of Being. So we needed to find new friends in the heavens.

Imagination, if not hope, springs eternal, and before long, fantasists and speculative fiction writers began spinning stories of civilizations inhabiting sister planets. Both H. G. Wells and Edgar Rice Burroughs portrayed humans meeting Martians. In *The War of the Worlds*, Wells described archetypal conquering aliens: the advanced, hostile, planet-hungry Martians descending upon Earth with total colonization in mind. Taking a much more "hail fellow, well met" tone, Burroughs, in his 11-book series,

depicted a Martian civilization in which humans could find comrades, at least if they were adventurers with a penchant for idealized feudalism—and skilled at fencing.

The historical search for life, then, initially held out hope for life on our sister planets. But the rapid advance of science quashed that hope. From looking for extraterrestrials on the Moon, on Mars, and on Venus, and imagining exotic alien cities with helpful automaton servitors and scantily clad residents, our expectations were turned around: Where once we saw not just life, but civilizations, we now saw only freezing, barren deserts, or gaseous, boiling hells. Luckily, things began looking up for extraterrestrial life in the 1990s.

LIFE IN EXTREMES—FROM SULFURIC SEA-WATER TO STELLAR DUST

"From the bottom of the ocean
To the mountains of the Moon"
—GRAHAM NASH, "CHICAGO" (1970)

In interstellar terms, what makes prime real estate? The characteristics of the good life, it seems, are pretty generalizable: a good home includes fences to keep out unwanted visitors, fresh air, sun and shade in equal parts, and plenty of clean water. That list pretty much sums up what astrobiologists initially thought were the basic requirements for life to evolve. As it turns out, that list may reflect at least in part the assumptions of a water-based life-form. Astronomers and astrobiologists use the phrase "the Goldilocks orbit" to sum up these characteristics. It is more formally known as the "habitable zone" (a term coined in 1959, but formalized in a 1992 paper by James Kasting). In our solar system, it refers to that band on the plane of orbit bounded near the Sun by the orbit of Venus, and at its farthest reach from the Sun, by the orbit of

WHERE ONCE WE SAW LIFE, WE NOW SAW ONLY FREEZING, BARREN DESERTS, OR GASEOUS, BOILING HELLS.

FACT OR FICTION?: The cover of 1929 science fiction magazine *Amazing Stories* shows the frightening image of invading Martian robots that occupied the imaginations of H. G. Wells and his readers. Jupiter provides a "shield" for the inner planets, dramatically illustrated by the Hoover effect it had on the Comet Shoemaker-Levy in 1994 (opposite, in a composite photo).

THAT'S LIFE: A 1977 discovery of hydrothermal vent communities and giant tube worms (left) "down there" on the ocean floor, feeding on sulfur-oxidizing bacteria, suggested forms of life "out there" in space that are equally "alien" to our usual concepts of life. Bacteria *Staphylothermus marinus* (below), another extremophile, is a hyperthermophilic (very high temperature loving) species found in deep ocean hydrothermal vents, thriving on volcanic sulfur and surviving in water temperatures of up to 98°C.

Mars—the best real estate in the system, where life could take hold and evolve.

Within the Goldilocks orbit, let's start with good fences—which Robert Frost would tell us make good neighbors. In this case, the good neighbor, Jupiter, actually makes the good fence. Jupiter, many scientists contend, enabled life to evolve on Earth because it Hoovers the solar system: Its great gravitational field sucks up stray space debris, which would otherwise find a way to the inner planets. In July 1994, Comet Shoemaker-Levy's spectacular collision with Jupiter illustrated exactly how good a fence it makes. So in looking for life elsewhere, planet-hunters would hope to find a gas giant guarding the yard of an Earth-size planet.

Now to the need for air, sun and shade, and water, which should be present in just the right amounts. Or should they? After all, biologists have known about anaerobic bacteria—which require no oxygen at all—since studies by Anton van Leeuwenhoek (1633–1723) and Lazzaro Spallanzani (1729–1799). In the 1970s, E. Imre

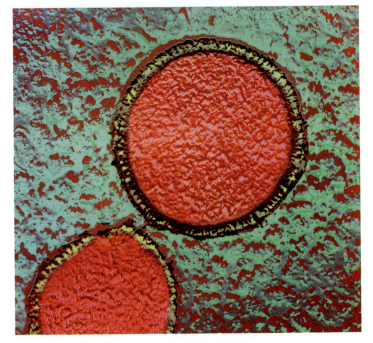

Friedmann found cyanobacteria growing in tiny crevasses between sandstone rock crystals in Antarctica's Dry Valleys, which everyone thought too dry and cold to support any kind of life. More recently, marine biologists exploring the ocean bottoms discovered colonies of tube worms and giant clams thriving in the dark, hot, toxic water around

WILD CARDS

Stalking the stellar surprises that, quite simply, change everything

FUTURISTS PROVOKE PEOPLE TO think past their assumptions with "wild cards": events currently considered unlikely but with the potential to transform everything. The movies *Armageddon* and *Deep Impact* both portray an asteroid-caused "extinction event," which is a classic example of a wild card. It is far more likely that any one of us would suffer an auto collision than that Earth would suffer an asteroid collision. But the car crash would impact only a small group of individuals; an asteroid impact would affect our civilization, our biosphere, perhaps even Earth's orbit.

Alien contact will be just such an event: currently considered unlikely—even if greatly anticipated—by many, it could potentially transform how we see ourselves, other life on this planet, our place in the universe, and our various relationships to higher spiritual powers. *Can* we estimate its likelihood and forecast its probable occurrence? The Drake Equation, first presented in 1961 by Dr. Frank Drake (now Chairman of the Board of the SETI Institute), seems to offer a starting point. Surely, if we can reliably estimate how many alien civilizations have evolved, should we not be able to estimate when we might hear from one of them? Unfortunately, the Drake Equation requires assessing seven preconditions for detecting a signal from an intelligent, extraterrestrial source, and then multiplying them. If any one term fails to pan out, the resulting answer is zero: zero probability of contact.

THE DRAKE EQUATION

$$N = R^* \times f_p \times n_e \times f_L \times f_i \times f_c \times L$$

Where,

N = the number of civilizations in the Milky Way Galaxy whose electromagnetic emissions are detectable

R^* = the rate of formation of stars suitable for the development of intelligent life

f_p = the fraction of those stars with planetary systems

n_e = the number of planets, per solar system, with an environment suitable for life

f_l = the fraction of suitable planets on which life actually appears

f_i = the fraction of life bearing planets on which intelligent life emerges

f_c = the fraction of civilizations that develop a technology that releases detectable signs of their existence into space

L = the length of time such civilizations release detectable signals into space

In an excellent exegesis of possible answers to Fermi, Stephen Webb, in his book *Where Is Everybody?* (quoting Fermi's question), has concluded that our evolutionary path was improbable and therefore probably unique—not in an "aren't we special?" sense, but rather in a "wow, what blind dumb luck" sense.

Nonetheless, he does point out that the issue is data-poor and that humans, in the absence of data, fall back on biases.

But the next 100 years will see our ability to scan the sky increase exponentially. And so, perhaps room exists for optimism regarding the probability of contact in the 21st century. Speaking of the new Allen Telescope Array, astronomer and SETI researcher Dr. Seth Shostak said, "This is the first instrument I think has a real chance of detecting a signal within our lifetimes." The first of many new instruments, and much more data: so hold on to that wild card for the next few decades—we may yet see it played.

DISCOVERIES ON EARTH HAVE LED SCIENTISTS TO WIDEN THE PARAMETERS FOR LIFE: THE GOLDILOCKS ORBIT IS EXPANDING.

deep sea thermal vents. How could they? They exist in symbiotic relationship with chemoautotrophic bacteria. These bacteria have no need for sunlight because they harness energy by metabolizing vented sulfur, rather than by photosynthesis. Another "extremophile," Chroococcidiopsis, is a primitive cyanobacterium that likes the underside of rocks, and has been found in extremely cold, extremely dry, and extremely salty environments, which makes it a candidate for the first step in terraforming Mars.

These discoveries on Earth have led scientists to widen the parameters necessary for life considerably: The Goldilocks orbit is expanding. Astrobiologists are growing cautiously optimistic about finding microbial forms of life even in freezing, barren deserts and gaseous, boiling hells. "The only thing that may stop life as we know [it] from taking root is lack of water, or very intense heat," says Ken Nealson of NASA's Jet Propulsion Laboratory (JPL). "Apart from that, in just about any place compatible with carbon chemistry, life figures out how to be there."

NEW MOON: Cassini will be the first spacecraft to orbit Saturn (below) in a joint mission of NASA, ESA, and the Italian Space Agency. The Huygens probe, aboard the Cassini, will be released to land on Titan, Saturn's largest moon (opposite), late in 2004, to gather data. Because Titan's surface composition is unknown, Huygens can land on either a solid or liquid surface.

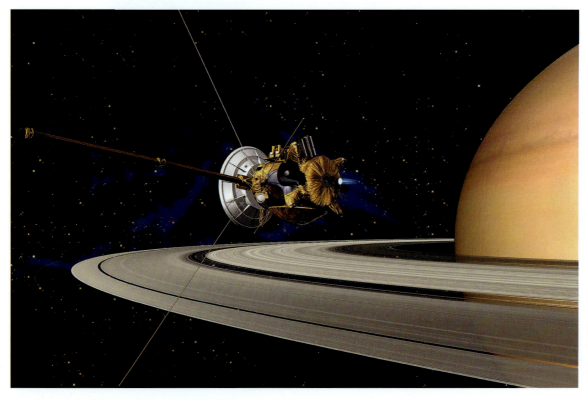

BEAM ME UP: A Mars Reconnaissance Orbiter (MRO, right) makes measurements and analyses of the planet's surface from above. Meanwhile, landers assess the surface, and probes, like the cryobot (opposite), their noses heated to melt the ice, descend through the sub-surface by using the planet's gravity. This should give an accurate analysis of the planet's makeup.

And as if to prove it, the world's space scientists have planned over 20 planetary missions between 2000 and 2015. The NASA/ESA Cassini orbiter was launched for Saturn in October 1997 and will drop ESA's Huygens probe on the moon Titan, after establishing an orbit around Saturn in July 2004. As of July 2003, four robotic probes were in flight to Mars.

For 2005 to 2030, the European Space Agency (ESA), Japan's NASDA, and NASA have proposed over a dozen projects. These include three missions each to the Moon and Mercury, two missions to Venus, one each to Pluto and Europa, and a further three to Mars. NASA plans for several of those to be "scout" missions, in which mobile explorers will collect and analyze samples from widely dispersed locations.

BEYOND THE SANDS OF MARS

But all these investigate the near neigh-borhood. What about the stars? A team at Pennsylvania State University, led by Alexander Wolszczan, was the first to find extrasolar planets—three of them—in 1990. Although dead worlds orbiting a dying star, they at least confirmed that ours was not the only solar system in existence with planets. In July 2003, a

THEY AT LEAST CONFIRMED THAT OURS WAS NOT THE ONLY SOLAR SYSTEM IN EXISTENCE WITH PLANETS.

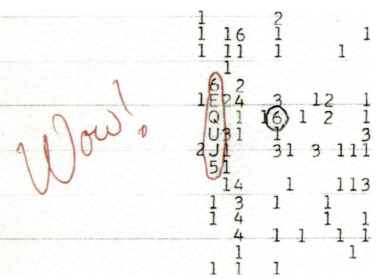

team led by Penn State's Steinn Sigurdsson found the oldest known planet: a gas giant 12.7 billion years old whose sun had been captured by a neutron star—the gas giant now orbits them both. As of July 2003, planet-hunters had found over 100 extrasolar planets, or exoplanets; they estimate that a potentially habitable Earth-like planet will be found within the next decade. Discovery of such a planet would amplify interest both in the search for extraterrestrial intelligence (SETI) and in mounting an interstellar exploratory expedition, if only by autonomous robotic probe. NASA already has an Interstellar Probe on the drawing board for 2015 or so, targeted to sail into the heliopause beyond our solar system's magnetic boundaries, about 90–110 Astronomical Units (AU) out. Success of that effort would pave the way for missions to the Oort Cloud or Alpha Centauri by 2050.

But space, as Douglas Adams (author of *The Hitchhiker's Guide to the Galaxy*) took such pains to tell us, is big. Searching thoroughly for both life and intelligence in our galaxy alone requires sorting through over 200 billion stars. To

OH WOW: The Big Ear radio telescope recorded the highest intensity values, 6EQUJ5—a SETI spike—provoking the observing scientist's response (above). The 2014 Terrestrial Planet Finder (right) will use linked telescopes to search for Earth-like planets.

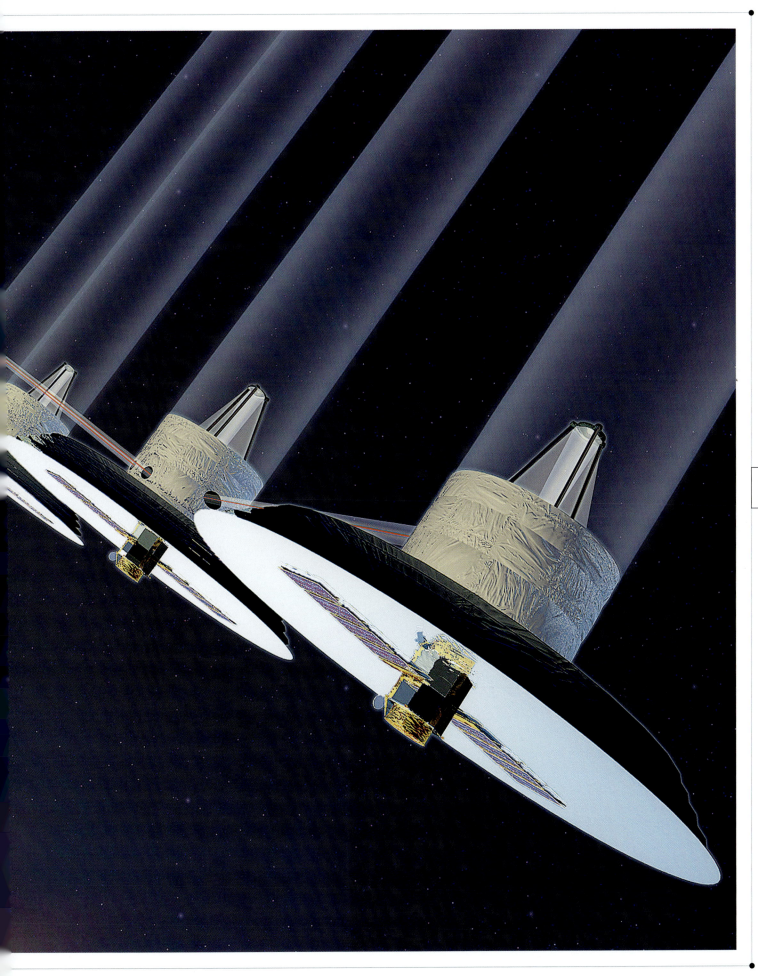

HEARING AIDS

New technology is vastly increasing our ability to hear what's going on in outer space

RADIO ASTRONOMER'S ZEN KOAN: What can be heard by one ear listening?

Nothing, compared to 350 listening in concert. By 2005, the new Allen Telescope Array (ATA), in Hat Creek, California, will spread 350 6.1-meter radio antennae across 10,000 square meters (1 hectare). Dedicated to SETI activities 24/7/365, it will speed the search for stellar messages a hundredfold.

As with optical telescopes, bigger is better with radio telescopes. Unfortunately, scaling up either eyes or ears reaches a threshold of clumsiness. The largest single-mirror telescope in the U.S. is Palomar's 5-meter Hale (a 6-meter mirror operates in Russia); much larger and the mirror's own weight would deform it. In making Hawaii's Keck Telescopes (I and II), designers opted to cast and polish 36 1.8-meter, hexagonal mirrors. Working in concert rather like a fly's compound eyes, these smaller mirrors provide magnification equivalent to a single 10-meter mirror—roughly double the size and quadruple the observing power of any existing optical telescope. Furthermore, with Keck I and Keck II targeting the same object, the combined data stream creates resolution equivalent to a telescope much larger than either of them singly.

So if we want to listen harder rather than see farther, what's the option? The largest single radio telescope is the 305-meter-wide dish at Arecibo; all effectively larger radio telescopes are arrays—fields of telescopes like steel sunflowers. Radio telescope arrays use computers to combine incoming signals via a process called interferometry. With this combinatorial technique, the degree of detail depends upon the dis-

tance between the telescopes. Because radio waves are between a thousand and several million times longer than light waves, radio telescopes must span larger diameters than optical telescopes to achieve equivalent resolution of detail.

For example, the Owens Valley Millimeter-Wavelength Array approximates the discriminatory power of the massive Arecibo installation. But it does so by combining data from the signal streams of six mobile, 10.4-meter radio telescopes deployed on a 500-meter, T-shaped railroad track. The 10 6.1-meter dishes of the Berkeley-Illinois-Maryland Array (BIMA), currently at Hat Creek, can be configured with baselines as large as two kilometers or as small as eight meters. By 2005, both the Owens Valley dishes and the BIMA dishes will move together to a new, high-altitude site and be renamed "CARMA"—Combined Array for Research in Millimeter-wave Astronomy. The resulting mob of radio telescopes will be the first heterogeneous array, featuring telescopes of

DISHING THE DATA: Dish antennae on radio telescopes at Owens Valley in California glow against the evening sky. They detect objects in the universe that emit radio waves of millimeter wavelengths, and they can act together to focus on a single object in the sky.

mixed sizes. The payoff? Combining the "small antennas' view of large regions of the sky with the large antennas' sensitivity to very faint objects" will produce "precise imaging of the Universe over a wide range of scales."

When completely deployed in 2005, the ATA's designers hope it will provide proof-of-concept for a proposed "Square Kilometer Array." One hundred times larger than the Allen Array, the Square Kilometer Array, if built, would have 10 times the collecting power of the Arecibo radio telescope. Bigger and cheaper—because these new arrays lower costs by using "off-the-shelf" dishes like those used for television reception.

Last item on the radio astronomer's wishlist? More computing power. Heightened data flow increases analytic indigestion: Bigger radio telescope arrays require bigger computational arrays. The 1999 SETI@home screensaver project at the University of California at Berkeley has produced an innovative solution to that problem. SETI@home encourages the public to participate by downloading cleverly designed "screensaver" software, and leaving their computers on, and on-line. During any desktop downtime, the screensaver links to a program at UC-Berkeley that distributes analytic assignments dynamically across this reservoir of computing resources. Why enlist an expanding pool of public participants? Because the Berkeley astronomers are evaluating output from Arecibo: looking for weak signals in a two-and-half-megahertz band of frequencies means analyzing 50 terabytes of data. With over four and a half million participants as of July 2003, they have completed the equivalent of one and a half million years of computing time, almost free.

Moore's Law—and advances in nanocomputing, optical computing, and quantum computing—will make the 21st century a data processor's paradise. This promises the tantalizing possibility of real-time aiming, repositioning, and computational analysis for radio telescopes. In the meantime, SETI@home has plans for data streams from a telescope in Australia, which watches a different arc of sky and, in the decade to come, may help process signal databases from installations like the Square Kilometer Array.

SEARCHING FOR LIFE AND INTELLIGENCE IN OUR GALAXY ALONE REQUIRES SORTING THROUGH OVER 200 BILLION STARS.

begin, as many as six planet-finding projects may be launched between 2004 and 2015. Three of them—COROT (Convection, Rotation, and Transit, coordinated by France's Centre National d'Etudes Spatiales, or CNES), Kepler (NASA), and Eddington (ESA)—would use optical instruments to search around the clock for planets in the habitable zone of relatively nearby stars. More complex mission designs include NASA's Space Interferometry Mission, slated to orbit the Sun starting 2009, or the 2011 Terrestrial Planet Finder, which would deploy an orbital telescope array to identify Earth-like planets and determine their suitability for life. The ESA plans to fly Darwin, which would orbit six small infrared telescopes between Mars and Jupiter to avoid the dust band between Earth and Mars, sometime around 2014.

In addition to looking for Earth-like planets, and sniffing out potential chemical precursors to life among interstellar gases using spectroscopy, radio astronomers also listen. The search by astronomers for extraterrestrial intelligence began in 1960 with astronomer Frank Drake's Project Ozma. Soviet scientists devoted the most radiotelescope time to SETI in the 1960s; in the 1970s,

NASA's Project Cyclops report spurred renewed efforts in the U.S., including the University of California's Project SERENDIP (Search for Extraterrestrial Radio Emissions from Nearby Developed Intelligent Populations), Ohio State's OSURO (Ohio State University

After furious fund-raising efforts, that project continued as a privately funded endeavor through the SETI Institute. Their targeted search effort, Project Phoenix, uses two of the world's largest radio telescopes—Arecibo and Jodrell Bank—to scan 1,000 nearby Sun-like stars

Radio Observatory, also known as the Big Ear Radio Observatory)—the guys that thought they'd spotted a signal and wrote "WOW!" on their output—and the Planetary Society's META project. NASA SETI funding started in 1988, formal observations began in 1992, and Congress cut funding in 1993.

intensively. They are turbo-charging this project with newly prototyped real-time signal detection equipment. In 2005, they will complement it with the Allen Telescope Array, which will search even more extensively. This 350-dish array will not only provide state-of-the-art radio astronomy research, but will be dedicated to

SETI research around the clock.

Other SETI projects include SERENDIP Arecibo, SERENDIP Australia, and Italian SERENDIP, which all piggyback searches with ongoing radiotelescope observations. SETI also double-checks visible frequencies. Optical SETI, represented by projects at Berkeley and Harvard as well as COSETI (Columbus Optical SETI, in the U.K.), search for messages that may be beamed our way using high-energy pulsed lasers.

What if contact occurs? What if we collect a message? Whatever do we say in reply? SETI designs billets-doux for interstellar friends. Earth's first interstellar message was beamed from Arecibo on November 16, 1974, aimed at globular cluster M13, 25,000 light-years away. Voyagers I and II also included scientific and cultural information. In 1999, Team Encounter, a Houston-based space company, transmitted a "Cosmic Call" to four stars more than 50 light-years away. It consisted of both scientifically composed and publicly penned messages (including little poems, famous quotes, personal hopes, and children's questions). Another "Cosmic Call" went out July 5, 2003, targetting five nearer stars. In 2006, Team Encounter will help private citizens greet interstellar neighbors in hard copy by launching an interstellar message capsule powered by solar sails.

Space scientists look somewhat askance at this casual commercialization of what may be the most significant human communicative act of this millennium. SETI's Director of Interstellar Message Composition, Doug Vakoch, has been coordinating both sides of humanity's brain by convening artists and scientists to discuss potential messages. One sticking point: How do we convey all of our most sterling qualities? First impressions can make or break a relationship. Mathematics may be the universal language, but equations and chemical symbols don't convey values well. Vakoch and his colleagues are struggling to encode messages of reciprocity and altruism, for example, by building up the concepts with images. Dutch mathematician Hans Freudenthal has suggested creating a Lingua Cosmica, Lincos, by stairstepping from simple mathematical relations through notions of time and space to systems relationships and human interactions. However we

WHAT IF CONTACT OCCURS? WHAT IF WE COLLECT A MESSAGE? WHATEVER DO WE SAY IN REPLY?

express ourselves, creating the ideaware for communication with extraterrestrials will be even more challenging than launching the hardware.

And the results will provoke more questions than answers. What might humanity's response be to intellectual companionship? Researchers compare the event to the shock occasioned in the 12th century when the Moors re-introduced the lost scientific knowledge of the Greeks and Romans into Europe via Spain. Intellectuals, religious leaders, and artists will feverishly attempt to integrate new perspectives, new knowledge, and new values into our various cultures. As in the 12th century, such an

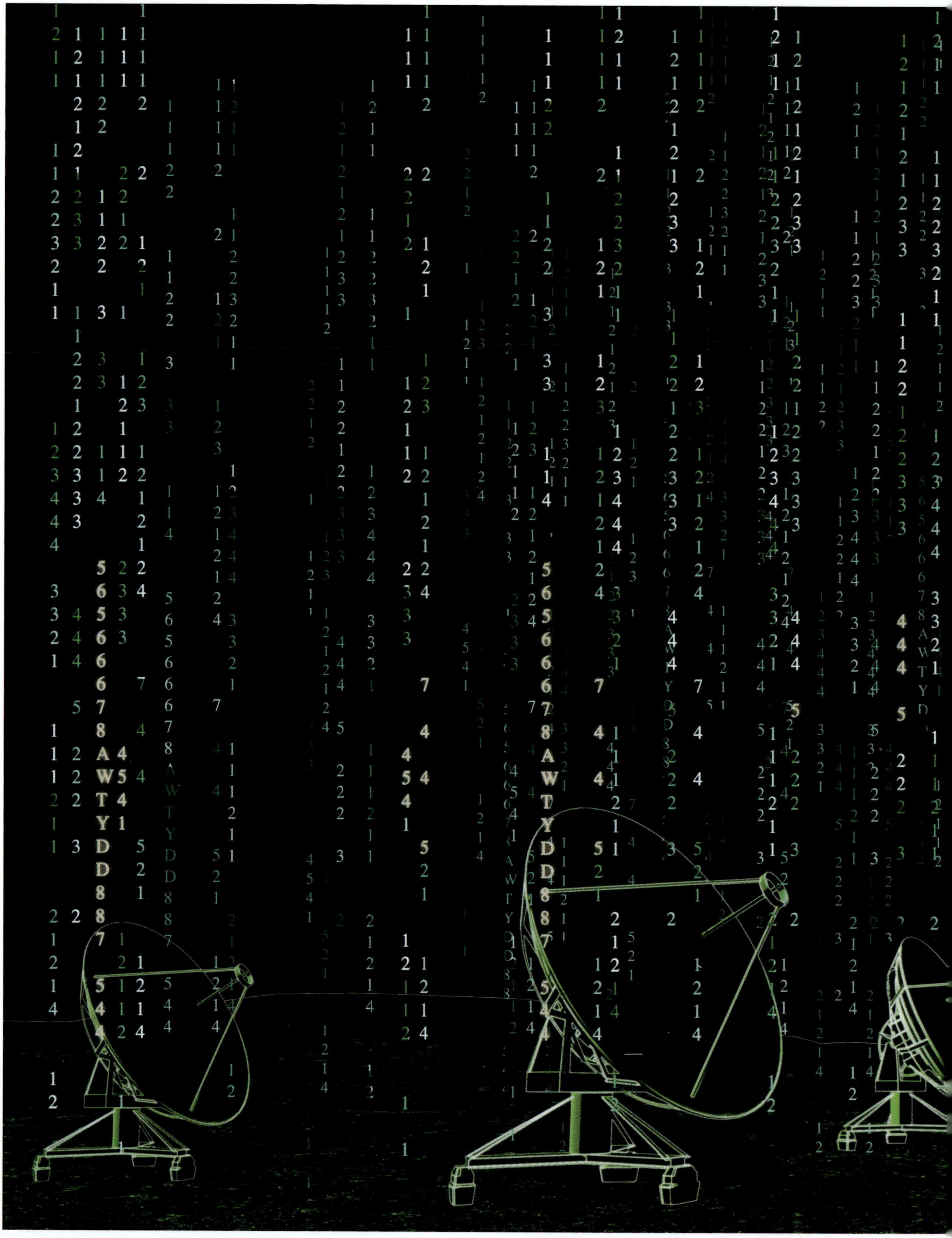

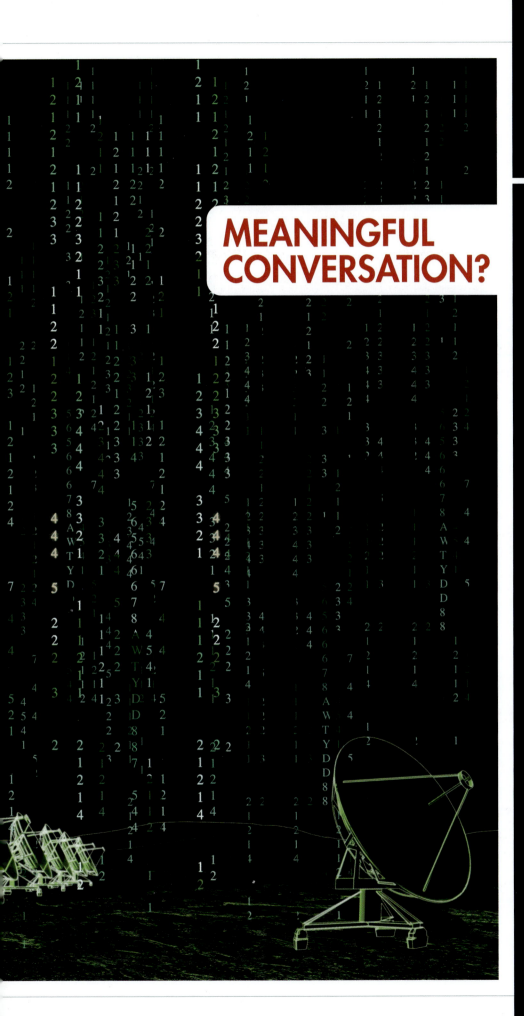

MEANINGFUL CONVERSATION?

As we sense signals raining down on us from outer space, we continually attempt to amplify and analyze them for some semblance of content and meaning—all in all, a semiotic specialist's dream or worst nightmare. Increasing the number of radio telescopes linked in an array enables us to hear more signals more strongly.

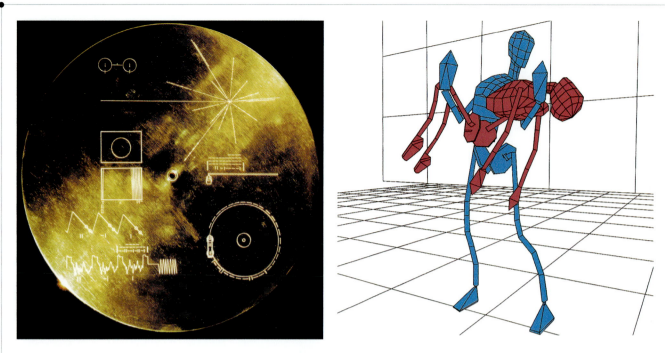

UNIVERSAL LANGUAGE: How does one communicate with aliens? The diagram above (left) illustrates how to play the "Sounds of Earth" recording aboard Voyagers 1 and 2; the image of figures (above, right) is an attempt by scientists like Doug Vakoch at the SETI Institute to show abstract human values like altruism; and the closeup of the plaque (below) aboard Pioneer 10 indicates the locations of Sun and Earth—our interstellar return address.

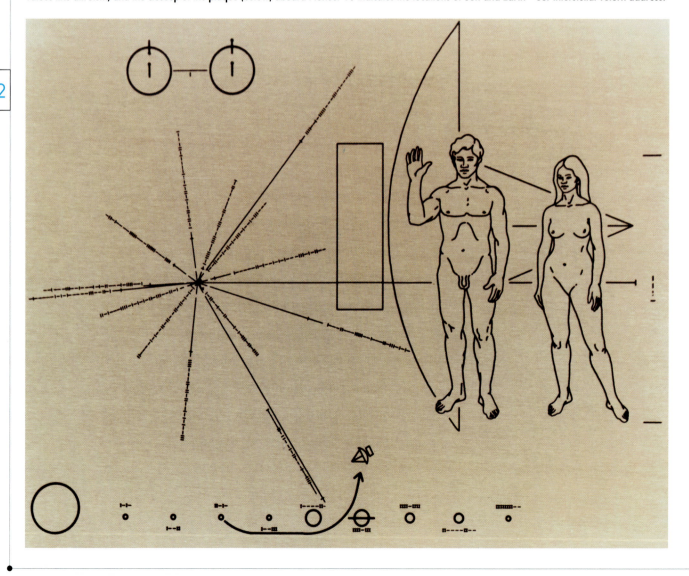

encounter could generate unparalleled creativity, not to mention revolutions in traditional institutions and worldviews. How would churches respond? Futurist and missionary Thomas Hoffman suggests that exomissiology (the study of religious missions to extraterrestrials) might be an appropriate future field of study in seminaries. In some distant century, will the Vatican fund an interstellar Jesuit mission, as suggested in Mary Doria Russell's *The Sparrow*?

BEYOND 2100: INFINITE POSSIBILITIES

Forecasting to 2100 requires a spirit of exploration; forecasting beyond that, an addiction to risk. To imagine human endeavors in space through the end of the 21st century and beyond, we must first question any assumptions we hold based on how the world works today. The nations that exist now may not exist tomorrow: The nation-state itself—a fairly recent invention—may become obsolete. Certainly our tools will evolve: smaller, faster, smarter, more inquisitive, more adaptive, self-repairing, and self-reproducing are all design characteristics under discussion now. The 21st century will assuredly see our mastery of genomic sciences, and with that, an emerging capacity to redesign ourselves.

Perhaps we are alone. If so, we won't be for long. Advances in nano-engineering and genomics will reinforce each other, and reinforce human efforts to jumar (a mountain-climbing term meaning to haul oneself up a nearly vertical slope) out of our gravity well. These changes will in turn amplify changes in us. New species of humans may well evolve out of 21st-century innovations—to paraphrase Marshall McLuhan, as we transform our tools, they transform us. We are generation zero of spacekind, and each successive generation will engender greater cultural and physical divisions

> # THE MORE OF US WHO SEARCH, THE MORE LIKELY WE ARE TO FIND SOMEONE, IF SOMEONE IS THERE TO BE FOUND.

from Earthkind. Over centuries and millennia we may find that the aliens with whom we most earnestly wish to communicate will be ourselves—or our evolved and sentient former tools.

Luckily we already have a plan: KEO, a UNESCO "Project of the Twenty-first Century." The KEO Project will create a time capsule, a digital letter with room for messages from all six-billion-plus of us, collected via mail and the Internet, digitally transcribed, and launched in 2006 into a 50,000-year orbit. Delivered to a distant generation, it will express our values, beliefs, stories, hopes, fears, and visions for humanity's future. If we cannot find a suitable conversational partner now, we can at least correspond with our future children.

But let us not give up hope prematurely. The more of us who search—public or private, human or robot—the more we are likely to find someone, if someone is there to be found. After all, you always find your missing book in the last place you look.

"There are two possibilities: Either we are alone in the universe or we are not. Both are equally terrifying."
—Arthur C. Clarke

PHOTOGRAPHY AND ILLUSTRATION CREDITS

COVER: Bob Sauls/John Frassanito & Associates

7, Bob Sauls/John Frassanito & Associates; 9, top, Hulton-Archive/Getty Images; bottom, left, Klaus Guldbrandsen/Photo Researchers; bottom, right, David Parker/Photo Researchers; 10, MSFC/NASA; 12-13, John Frassanito & Associates; 14, Courtesy of The Planetary Society; 15, JPL/NASA; 16-17, John Frassanito & Associates; 18, JPL/NASA; 19, GRC/NASA; 20-21, John Frassanito & Associates; 23, Reuters NewsMedia/Corbis; 24, left, top, Courtesy of International Space University; left, bottom, Xinhua/Sovfoto; right, Xinhua/Sovfoto; 26, top, XCOR Aerospace; bottom, Space Adventures; 28-29, John Frassanito & Associates; 30-31, Bob Sauls/John Frassanito & Associates; 32-33, John Frassanito & Associates; 34, Cory Bird/Scaled Composites; 35, GSFC/NASA; 36, MSFC/NASA; 39, John Frassanito & Associates; 40, JSC/NASA; 42-43, John Frassanito & Associates; 45, JPL/NASA; 46, Lowell Observatory; 47, Detlev van Ravenswaay/Photo Researchers; 48-49, JPL/NASA; 48, JPL/Malin Space Science Systems/NASA; 50, NASA/Photo Researchers; 51, JPL/U.S. Geologic Survey/NASA; 52, JPL/Malin Space Science Systems/NASA; 53, left, LRC/NASA; right, JPL/NASA; 54, Denman Productions/ESA; 56, Extreme Ultraviolet Imaging Telescope Consortium/JPL/NASA; 57, John Frassanito & Associates; 58-59, Bob Sauls/John Frassanito & Associates; 60, John Frassanito & Associates; 62-63, David A. Hardy/Photo Researchers; 64-65, John Frassanito & Associates; 67, Corbis; 68, MSFC/NASA; 70, MSFC/NASA; 72-73, Scaled Composites/X Prize Foundation; 72, Canadian Arrow; 73, Tethers Unlimited; 74, NASA; 74-75, MSFC/NASA; 76, both, John Frassanito & Associates; 78, GRC/NASA; 79, top, GRC/NASA; bottom, JPL/NASA; 80-81, John Frassanito & Associates; 82-83, Bob Sauls/John Frassanito & Associates; 84, MSFC/NASA; 85, John Frassanito & Associates; 86-87, Johns Hopkins University Applied Physics Laboratory/Southwest Research Institute; 89, Bettmann/Corbis; 90, JPL/NASA; 91, JSC/NASA; 92, JPL/NASA; 93, ARC/NASA; 94-95, John Frassanito & Associates; 95, JPL/NASA; 96, JPL/NASA; 97, Chris Butler/Photo Researchers; 98-99, LRC/NASA; 100, John Frassanito & Associates; 101, JPL/NASA; 102-103, Bob Sauls/John Frassanito & Associates; 104, JPL/NASA; 105, JPL/NASA; 106-107, all, John Frassanito & Associates; 108-109, John Frassanito & Associates; 111, Space Adventures; 112, TGV Rockets; 114, Dominic Hart/ARC/NASA; 115, ARC/NASA; 116, left, Alfred Pasieka/Photo Researchers; bottom, Dr. Peter Harris/Photo Researchers; 117, AP/Wide-World Photos; 118, JPL/NASA; 119, JPL/NASA; 120-121, JSC/NASA; 121, NASA; 122, NASA; 123, JSC/NASA; 124-125, John Frassanito & Associates; 125, MSFC/NASA; 126-127, Bob Sauls/John Frassanito & Associates; 128-129, both, Sasakawa International Center for Space Architecture (SICSA)/University of Houston; 130-131, MSFC/NASA; 133, John Frassanito & Associates; 135, John Frassanito & Associates; 136, Courtesy of The Mars Society; 137, MSFC/NASA; 138-139, both, John Frassanito & Associates; 140, NASA, 141, Craig Hammell/Corbis; 142, top, JSC/NASA; bottom, John Frassanito & Associates; 143, Photodisc/Getty Images; 144, NASA; 144-145, John Frassanito & Associates; 146, NASA; 148-149, Bob Sauls/John Frassanito & Associates; 151, Julian Baum/Photo Researchers; 152-153, JPL/NASA; 155, Lowell Observatory; 156, Space Telescope Science Institute/JPL/NASA; 157, Bettmann/Corbis; 158, top, D. Foster/Woods Hole Oceanographic Institution; bottom, Wolfgang Baumeister/Photo Researchers; 160, AFP/Corbis; 161, Detlev van Ravenswaay/Photo Researchers; 162, NASA; 163, JPL/NASA; 164, Jerry Ehman/Big Ear Radio Observatory Website; 164-165, JPL/NASA; 166, David Nunuk/Photo Researchers; 168, David Parker/Photo Researchers; 170-171, Bob Sauls/John Frassanito & Associates; 172, top, left, JPL/NASA; top, right, Courtesy of Dr. Doug Vakoch/SETI Institute; bottom, Bettmann/Corbis.

INDEX